Immunoprophylaxis of Infectious Poultry Diseases

The Authors

Dr **Raghvendra Pratap Singh** graduated (B.V.Sc & A.H, Hons) from GBPUA & T, Pantnagar and Postgraduated from I.V.R.I, Izatnagar in the discipline of Veterinary Public Health with Veterinary Immunology. He has published two research papers in national Journal of repute. Dr Singh is currently working as Touring Veterinary Officer at Indara, Mau, Govt. Bihar. Apart from academics, excelled in sports and games during stay at GBPUA & T, Pantnagar as captain, University volleyball team and Vice-Captain, University football team.

Dr **Vibha Yadav** obtained B.V.Sc. & A.H. degree and M.V.Sc. in Veterinary Microbiology with Chancellor Gold Medal from College of Veterinary Science, N.D. University of Agriculture and Technology, Kumarganj, Faizabad. Currently she is working as an Assistant Professor in the Department of Veterinary Microbiology, N.D. University of Agriculture and Technology, Kumarganj, Faizabad. Dr Vibha also worked as Veterinary Officer in Animal Husbandry Department, State Government of Uttar Pradesh at State Hospital and Canine Rabies Control Unit and also worked as Nodal Officer at Divisional Animal Laboratory, Faizabad. Her area of research interest is Veterinary Bacteriology. She has been recipient of many awards and honours viz Excellence in Teaching and Young Scientist awards. She has published eighteen research papers/review articles/short communications/popular articles/hindi articles/pumplets in journals of repute.

Immunoprophylaxis of Infectious Poultry Diseases

Dr Raghvendra Pratap Singh

B.V.Sc. and A.H. (Hons.), G.B.P.U.A. and T., Pantnagar, (U.K.)
M.V.Sc. (Veterinary Public Health and Veterinary Immunology),
I.V.R.I., Izatnagar, (U.P.)
J.R.F., N.E.T., M.I.A.V.P.H.S.
Veterinary Officer, Department of Animal Husbandry,
Government of Uttar Pradesh
at Indara, Mau

Dr (Smt.) Vibha Yadav

B.V.Sc. and A.H. (Hons.), M.V.Sc. (Gold Medalist)
Asstt. Professor, Department of Veterinary Microbiology
College of Veterinary Sciences and Animal Husbandry
N.D.U.A. and T., Kumarganj, Faizabad, (U.P.)

2018

Daya Publishing House®
A Division of

Astral International Pvt. Ltd.
New Delhi – 110 002

Published by   :   **Daya Publishing House®**
                        A Division of
                        Astral International Pvt. Ltd.
                        – ISO 9001:2015 Certified Company –
                        4736/23, Ansari Road, Darya Ganj
                        New Delhi-110 002
                        Ph. 011-43549197, 23278134
                        E-mail: info@astralint.com
                        Website: www.astralint.com

Digitally Printed at   :   **Replika Press Pvt. Ltd.**

Dedicated to My Father

Dr. Lakshmi Narayan Singh

*The Great Inspiration to
Serve Poors*

Acknowledgements

The authors of this book cordially acknowledge for granting the permission of its publication by *Dr. Charan Singh Yadav*, Hon'ble Director (Administration and Development), Department of Animal Husbandry, Govt. of Uttar Pradesh and for kind valuable suggestions after its thorough critical revision by *Dr. J. M. Kataria*, Hon'ble Director, I.C.A.R. - C.A.R.I., Izatnagar, U.P.

Dr. Raghvendra Pratap Singh
Dr. (Smt.) Vibha Yadav

Preface

The Indian agriculture sector economy has a wide contribution from poultry sector with approximatly 1 per cent of GDP. As per APEDA-Agri Xchange 2016, India ranks 3rd in egg production after China and USA while 4th in broiler production after China, Brazil and USA in the world. However, per capita availability of required number of eggs and amount of poultry meat is still low in the country, which needs more production from lower inputs. In this context, disease control in poultry farms becomes more vital in getting maximum/optimum production. Major poultry diseases not only effect the production badly but also increase the cost of production in terms of loss of money in treatments of the diseases. Prevention is always better than treatment. So, immunoprophylaxis in poultry production and disease management is the best choice now-a-days. A hygienic and profitable poultry farming cannot be imagined without correct vaccination of the flocks being reared at the farms. Immunoprophylaxis is the backbone of profitable poultry farming. It has been long felt need for such a book of its kinds which will be very useful for all poultry veterinarians, veterinary students, teachers and poultry professionals in this context. This book will certainly be more useful to the veterinarians/nodal officers of commercial layer and breeder poultry units established and being established under the policy of government of Uttar Pradesh and other states. The book contains concise information on major disease management through applied poultry vaccination and also practical oriented depth of knowledge in the field of poultry vaccinology. The book is immensely helpful in understanding the formulation of vaccination schedules for broilers, layers and breeders. The present book not only covers the preventive aspects of diseases but also the all-round information dealing with theoretical and applied practical aspects of poultry immunology and vaccination. The language of the book is very simple, learnable and understandable for not only students and teachers of the subjects of poultry production and management, poultry pathology, poultry immunology and poultry preventive medicine but also for poultry professionals.

This book will definitely play and serve a critical role in imparting knowledge in poultry egg and poultry meat industry in terms of disease free maximum/ optimum production of good quality eggs and meat from poultry for human consumption. Although the manuscript has been typed very carefully, there may be some typographical errors. So, your suggestions and critics about this first edition of the book are thankfully invited to further improve the contents and materials in this presented edition to the E-mail: avadhvanshi.raghvendra@gmail.com.

My sincere thanks to extend help from my beloved family members, Smt. Sangeeta Rani Singh (wife), Meenakshi Singh (daughter) and Love Pratap Singh (son) during preparation of this book. Thanks to Dr (Smt.) Vibha Yadav for co-editing of this book. Extended thanks to Dr. Abhishek Vishwas, Senior Scientist, I.C.A.R. – C.A.R.I., Izatnagar for his help and valuable suggestions for improving the text.

We hope, this present edition of the book will definitely add in scientific knowledge and skill in poultry vaccinology not only to be applied in the field for improving the production and management associated with infectious poultry diseases to produce optimum and safe poultry products like eggs and meat intended for human consumption but also help the students, teachers and poultry professionals to understand the basics of applied poultry vaccinology in simple way.

Dr. Raghvendra Pratap Singh
Dr. (Smt.) Vibha Yadav

Contents

Chapter 1

Reovirus Infections

Reovirus infections of poultry are widespread in all commercial flocks. The virus causes a variety of disease conditions in poultry like viral arthritis, tenosynovitis, femur head necrosis, brittle bone disease, runting-stunting syndrome and malabsorption syndrome. Viral arthritis/tenosynovitis is chiefly a disease of meat-type (broilers) poultry resulting in to leg weakness because of swelling of one or both hock (tibio-tarsal and tarso-metatarsal) joints which are the main load bearing joints in birds causing acute lameness. However, this condition is not common up to 4-5 weeks aged birds but its peak incidence is generally seen at 7th onwards weeks of age. Occasionally broiler breeders are affected during peak production. Morbidity is about less than 10 per cent and mortality is very low.

Affected joints are swollen and inflamed. In more severe cases rupture of gastrocnemious tendon and erosion of articular cartilage occur. Occasionally, one or more digital flexor tendons are ruptured. Rupture of gastrocnemious tendon is accompanied by hemorrhages which in turn causation of green discoloration of the skin at the concerned joints.

Avian reovirus is also associated with cloacal pasting and mortality (Dutta and Pomeroy, 1969), ulcerative enteritis (Krauss and Ueberschar, 1966), enteric disease (Dutta and Pomeroy, 1969), respiratory disease (Fahey and Crawley, 1954), heart lesions in young broilers (Bains *et al.*, 1974), lesions in kidney and liver (Bagust and Westbury, 1975), malabsorption and brittle bone disease (Goodwin, *et al.*, 1993).

Reovirus is also isolated from turkeys with tenosynovitis (Page *et al.*, 1982) and muscovy ducks (McNulty, 1993). A recent review of molecular virology of avian reoviruses can be found (Kaleta and Heffels-Redmann, 1996). Avian reovirus belong to genus Orthoreovirus in the family reoviridae, measuring 70-80 nm with icosahedral symmetry, non-enveloped with double shelled arrangement of surface proteins. The genome is ds-RNA with 10 segments in it. Out of 11 proteins synthesized, 9 are structural proteins and 2 are non-structural proteins. Protein coding assignments of all the 10 genome segments of strain S-1133 have been determined (Varela and Benavente, 1994). The strain S-1133, isolated in U.S.A. is the basis of many commercial vaccines and appears widespread throughout the world, although many regional variants exist.

Avian reoviruses are stable between pH 3.0 and 9.0 but inactivated at 56°C temperature in less than 1 hour time period. The virus can survive for up to 1o days on feathers, wood shavings, glass, rubber and galvanized metals for 10 weeks in water with limited effect on infectivity. Avian reovirus is relatively resistant to disinfectant like 2 per cent formaldehyde at 4°C temperature (Meulemanns and Halen, 1982) but 100 per cent ethyl alcohol is effective (Petek *et al.*, 1967).

Epidemiology

Both vertical and horizontal transmissions of avian reoviruses have been recognized. Vertical transmission through eggs has been confirmed experimentally (Al-Mufarrej *et al.*, 1996, Menendez *et al.*, 1975). Congenitally infected chicks serve as infection nuclei for other hatch mates but mostly infected via oro-fecal route (Jones and Onunkwo, 1978). Infection via respiratory tract is also noted. Reovirus may also enter through abraded skin and established in hock joints infected from contaminated litter (Al-Afaleq and Jones, 1990).

Although it is a disease of heavy meat type birds, but sometimes light egg layers may also be affected (Schwartz *et al.*, 1976). Resistance to reovirus infection in chicken is age linked. In chicks infected at day-old, higher intestinal virus titer and more sever joint lesions developed than in those infected when older (Montgomery *et al.*, 1986, Roessler and Rosenberger, 1989). Infectious agents which enhance the effect of reovirus pathogenesis in the joints of chicken include Mycoplasma synoviae (Bradbury and Garuti, 1978) Staphylococcus aureus (Kilbenge and Wilcox, 1983), IBD (infectious bursal disease) virus (Moradian *et al.*, 1985) and CIA (chicken infectious anemia) virus (McNeilly *et al.*, 1995).

Pathogenesis

Vertical transmission of reovirus usually occurs at a low rate and most of the chicks of a flock become infected at an early age via oral or occasionally through respiratory route from the small nuclei of congenitally infected hatch mates or from the environment (Al-Mufarrej *et al.*, 1996). Experimental infection of adult S.P.F. (specific pathogen free) hens via nasal, tracheal or oesophageal routes, showed distribution of virus to all the areas of the respiratory, enteric and reproductive tracts and the tendons of hock joints (Menendez *et al.*, 1975). After 30 hours of infection, viraemia occurs. Despite of widespread tissue dissemination, the principal site of viral replication is enteric tract (Kibenge *et al.*, 1985).

Epithelial cells of small intestine and Bursa of Fabriceus are the main sites of primary infection and portal of entry of the virus which rapidly spreads to other organs within 24 to 48 hours of infection. But the important sites of consequences of viral replication are seen in tibiotarsal–tarsometatarsal (hock) joints (Jones *et al.*, 1989). It leads to joint damage and even rupture of tendons. The other target organ is liver resulting in to hepatitis (Jones and Guneratne, 1984). The tissue tropism of avian reovirus is genetically determined (Meanger, *et al.*, 1999).

Early indications of effects on the joints include soft swelling of the joints involving synovial membranes and surrounding tissues. The synovial fluid may become turbid because of secondary bacterial or Mycoplasma infections. As the disease progresses, petechiae may be seen in the synovial membranes and small erosions on the articular cartilages of the hock joints. Adhesions between the tendons and fibrosis of the tissues prevent smooth movement and shanks may be swollen when digital flexor tendons are affected. In older and heavier birds, the gastrocnemius tendons and occasionally the digital flexor tendons may rupture and it leads to lameness, reluctant to move for food and water, ultimately leading to death due to starvation and dehydration.

Lesions

Histopathological changes include thickening of tendon sheaths due to edema, hypertrophy and hyperplasia of the synoviocytes, villous proliferation of the synovial membranes and invasion with inflammatory cells. Later on, the loose connective tissues around the tendon sheaths are replaced by fibrous tissues. The histological changes due to reovirus are characterized by diffused lymphocytic inflammation, while those caused by Staphylococci are focal purulent synovitis.

Microscopic lesions related to hepatitis, spleenitis, bursitis, arthritis, pericarditis and myocarditis may be seen in this disease (Kerr and Olson, 1969).

Signs and Symptoms

Lameness is the most common manifestation of this disease in poultry. The clinical signs are not clearly pathognomonic and may resemble to those caused by other aetiological agents like Mycoplasma synoviae and Staphylococcus aureus. The disease primarily affects meat type of birds *i.e.* broilers and occasionally light egg laying breeds of chickens (Schwartz *et al.*, 1976).

Diagnosis

Diagnosis is based on clinical signs and symptoms followed by laboratory confirmations. Laboratory confirmation is achieved by demonstration of virus/ isolation and identification of virus by traditional as well as by novel methods. Examination of sick as well as healthy birds should be done. Selected specimens collected aseptically can be sent to laboratory in separate containers. The specimen may include feces, trachea, liver, bursa, kidney, spleen, hypo tarsal sesamoid bone including associated tendons, hock articular cartilage and synovial membrane. Swabbing of joints may also be done (Jones and Georgiou, 1985). Specimens should be sent to laboratory in suitable transport medium. If delay occurs in processing, specimens can be stored temporarily at 4°C temperature or for longer periods at -20°C or temperature below it.

Isolation of reovirus is achieved by inoculation of specimen material in to fertile chicken eggs or chick embryo cell cultures. Embryonating eggs, preferably from

SPF flocks are inoculated via yolk sac route after 6 days of incubation. Virulent reoviruses typically kill the embryos with in 5-6 days of inoculation and embryo appears hemorrhagic with necrotic lesions on liver.

Inoculation of cell cultures with reovirus results in syncytium formation in the cell sheet with affected cells lifting off in to the medium after a few days. In H and E (Hematoxylin and Eosin) staining, eosinophilic intranuclear inclusion bodies are seen in the affected cells. Isolation of reovirus from the hock joints may be considered as diagnostic. However, isolation and identification of reovirus from the tissues is time-consuming, so other more rapid methods can be employed.

The rapid methods may include dot-blot hybridization (Yin and Lee, 1998), P.C.R./Polymerase Chain Reaction (Xie *et al.*, 1997) and P.C.R. combined with R.F.L.P./Restriction Fragment Length Polymorphism (Lee *et al.*, 1998).

In immunological methods/tests to detect antibodies against avian reovirus, A.G.I.D. (Agar Gel Immunodiffusion), V.N.T. (Virus Neutralization Test), I.I.F. (Indirect Immunofluorescence) and E.L.I.S.A. (Enzyme Linked Immunosorbent Assay) are of greater importance. Additionally, western blot method has also been described (Endo-Munoz, 1990). However, virus isolation method is considered as "gold standard" for avian reovirus diagnosis. But, the use of P.C.R.-R.F.L.P. combined methodology appears to show promise for tracing the source of infections.

Prevention and Control

Reovirus is relatively resistant out side the host. Vertical transmission of the virus puts another challenge. Thus, the main approach to reovirus control is vaccination using live and or killed vaccines. Since the chicks are most susceptible to reovirus infection immediately after hatching (Jones and Georgiou, 1984), vaccination protocols are designed to protect these chicks during the early days of life. This has been accomplished by passive immunity from MAbs (Maternally derived Antibodies) following vaccination of the breeder hens or by active immunity after early vaccination with a live vaccine. Importation of stock or eggs is not advised from any region where the disease caused by a particularly virulent reovirus. Reovirus is relatively resistant and survives well outside the host on egg shells, egg boxes and other fomites. Poultry products remain normally safe unless contaminated with fecal materials from gut.

Immunological Considerations

In the reovirus infection, immunity involves B-cells and T-cells (both systems) with the B-cell system being more important in protection. Maternally derived antibodies (MAbs) to reovirus are effective in protecting chicks infected at day-old (Van der Heide *et al.*, 1976). The protective effect conferred by MAbs is the basis of breeder vaccination. Initially, Ig (Immunoglobulin) A has protective role against infection of reovirus entered through oral route. T-lymphocytes and plasma cells are the predominant inflammatory cells in the synovium. In the acute phase, T-cells (CD8+) remain present in low numbers but most activity are seen in sub acute phase with increased number of CD4+ and CD8+ cells. Aggregates of T-cells, Ig M positive B-cells and plasma cells are also present. While the chronic stage is characterized by chiefly large number of CD4+ T-cells with few Ig M positive B-cells. It has been also suggested that avian reovirus arthritis is an acute-immune disease which can be a model of research for rheumatoid arthritis in humans, although no rheumatoid factor has been demonstrated (Marquardt *et al.*, 1983).

The chicks are most susceptible to avian reovirus infection immediately after hatching (Jones and Georgiou, 1984). So, the vaccine protocols are designed to protect these chicks during the early age of life. However, in general the use of live vaccines in chicks at one-day-old has not been very successful. This may be related to the poor intestinal immunity in very young chicks after immunization at this stage (Mukiibi–Muka, 1997).

Later on, efforts were made towards live or inactivated vaccines for breeding stock to provide passive immunity in the progeny chicks via egg yolk (Van der Heide *et al.*, 1976). Inactivated preparations from strain "s 1133" induced MAbs which were relatively short lived. Later on a preparation of "s 1133" was used which was attenuated after 74 embryo passages to vaccinate broiler breeders between 10 to 15 weeks of age through drinking water. The progeny were subsequently found to be resistant to oral and subcutaneous challenge with homologous virus but not against different serotypes (Rau *et al.*, 1980). The above vaccine was found to reduce the incidences of lesions in hock joints of progeny after challenge at one-day-old age. The use of live vaccine as primer early in life, followed by an inactivated vaccine given at 6th week of age and later on prior to start of laying (Giambrone, 1985) was found more promising in development of immunity. More recent developments have been involved as the use of coarse spray administration of a cell cultured clone of strain "s 1133/66" (Giambrone and Hathcock, 1991). This preparation resulted in higher antibody levels than egg-passaged vaccine. Inactivated reovirus vaccines are frequently administered to breeder flocks in combination with other killed preparations against ND and EDS–76 virus infections. Some breeders vaccinate parent flocks with killed vaccine before exporting eggs or chicks so that good levels of MAbs could protect the chicks during the post-hatch period when the chicks are most susceptible. MAbs generated by a conventional reovirus vaccine, mostly based on the "S 1133" strain from USA, may not be protective against the antigenic variants which exist in some other countries.

Vaccination

The purpose of vaccination is:

1. To protect growing pullets and breeders from tenosynovitis.
2. To promote the development of high levels of MAbs in breeders to be transferred to the progeny during the egg laying period.
3. To prevent leg problems and poor growth performance associated with reovirus infections in broilers.

To achieve the above purposes of quality vaccination, generally s 1133, UMI 203, 2408, 1733, CO 8, 3005 and ss 412 reovirus strains are used in vaccines. Out of these strains, s 1133 is available for both live and inactivated, UMI 203 is available for live only and rest are available for inactivated vaccines. Strain s 1133 and UMI

203 are related with tenosynovitis and malabsorption syndrome, CO 8 is related to malabsorption syndrome only, 3005 is related to malabsorption syndrome, femoral head necrosis and brittle bone disease while ss 412 strain is related with malabsorption syndrome and proventriculitis. s 1133 is a strain commonly used in live vaccines which has been originally isolated from flocks with tenosynovitis. Live reovirus vaccines strains are highly attenuated in the laboratory. Less attenuated vaccine strains are shed vertically to progeny and may persist in tendons resulting in to tissue damage. Other vaccine strains which are utilized in inactivated vaccines are specific for associated syndromes.

Breeder Reovirus Vaccination

Programme – 1

This live vaccination programme is for use in breeder flocks to protect pullets from tenosynovitis and is only used when low MAbs levels are needed in progeny. Young birds are most susceptible to the pathogenic effects of reovirus.

Live reovirus vaccines are generally administered by subcutaneous route or by wing web stab. Drinking water administered reovirus vaccines have been found too pathogenic for use in breeders. s 1133 highly attenuated reovirus strain vaccine is used in younger breeder birds but s 1133 less attenuated reovirus strain vaccine is used in older breeder birds. In high field challenge areas, a highly attenuated live tailor made vaccine is used to vaccinate the flocks as early vaccination between 7 days to 3 weeks old aged birds. Multiple vaccinations are needed to maintain an adequate level of immunity to prevent tenosynovitis in the pullets. A highly attenuated strain is designed for younger birds (<3 weeks of age) which can also be used for revaccination when breed sensitivity to less attenuated vaccine strains exists. However, live reovirus vaccines intended for use in older birds may be pathogenic for younger birds.

Programme – 2

The objective of this live/inactivated vaccination programme is to protect breeder flocks from tenosynovitis and development of uniform and high levels of MAbs in the progeny chicks.

In this programme, firstly live vaccine is used to prime (prepare) the breeder's immune system to obtain the optimum immune response after vaccination with inactivated vaccines. One to three live vaccines are to be given depending upon the severity of the field challenge. Birds which are successfully primed with live vaccines respond the best to inactivated reovirus vaccine administered thereafter.

Inactivated reovirus vaccine is generally given 4 weeks prior to the onset of egg production (at 12th week of age of birds) through intramuscular (i/m) or subcutaneous (s/c) route to allow sufficient time for the development of full immune response. There are some inactivated reovirus vaccines also containing certain strains which stimulate production of antibodies specific for other reoviruses induced diseases such as malabsorption syndrome, brittle bone disease and femoral head necrosis.

A second inactivated reovirus vaccine is given at 40–45 weeks of age to maintain high and uniform levels of hen antibody titers throughout the later part of the egg production period. The decision to use this midlay vaccination should be based on the assessment of hen flocks titers at 35–40 weeks of age.

However, malabsorption syndrome, brittle bone disease and femoral head necrosis may be caused by other etiological agents than reovirus. The stress of handling during lay period for vaccination may cause decrease in egg production. The residue and tissue reaction from vaccination with oil-adjuvented inactivated vaccines may be detectable for many weeks post-vaccination.

Programme – 3

The objective of this live/inactivated vaccination programme is to protect the breeders from tenosynovitis and development of high and uniform MAbs levels in the progeny chicks.

An alternative to the midlay inactivated vaccination is to administer two inactivated vaccines prior to the onset of egg production. The use of two inactivated reovirus vaccines makes possible to include additional strains of virus in vaccines for this programme. These strains broaden the breeder's antibody response to improve the MAbs protection of the progeny chicks against malbsorption syndrome, femoral head necrosis, brittle bone disease and tenosynovitis. To produce an optimum antibody response, two inactivated reovirus vaccines should be administered at least 4 weeks apart *i.e.* 1st vaccination at 8th week of age and 2nd vaccination at 12th week of age (4 weeks prior to the onset of egg production) of birds through i/m or s/c route allowing for full immune response development up to optimum economical production life. Prior to these inactivated vaccinations, live vaccination priming is recommended between 1st-3rd week of age. The birds primed with live vaccines respond better to inactivated vaccines.

Broiler Reovirus Vaccination

The objective is to use live vaccination programme in broiler flocks having low MAbs level. This programme has been reported to be successful in improving broiler performance when used in low MAbs titer chicks.

MAbs provide passive protection to the broiler chicks for the first 2 weeks of life, but its effect on live vaccination is not clearly understood. These vaccines are usually given subcutaneously at the hatchery in combination with MD (Marek's Disease) vaccine. However, simultaneous administration of live reovirus and MD vaccine can produce interference in development of optimum immunity against Marek's disease. This potential of interference becomes more when diluted MD vaccines are used with reovirus vaccine.

REFERENCES

Al-Afaleq, A. I. and Jones, R. C. (1990). Localisation of avian reovirus in the hock joints of chicks after entry through broken skin. *Res. Vet. Sci.*, 48, 381-382.

Al-Mufarrej, S. I., Savage, C. E. and Jones R. C. (1996). – Egg transmission of avian reoviruses in chickens: comparison of a trypsin-sensitive and a trypsin-resistant strain. *Avian Pathol.*, 25 (3), 469-480.

Bagust, T. J. and Westbury, H. A. (1975). Isolation of reoviruses associated with diseases of chickens in Victoria. *Aust. Vet.*, 51, 406-407.

Bains, B. S., Mackenzie, M. and Spradbrow, P. B. (1974). Reoviruses associated with mortality in broiler chickens. *Avian Dis.*, 18,472-476.

Bradbury, J. M. and Garuti, A. (1978). Dual infection with Mycoplasma synoviae and a tenosynovitis-inducing reovirus in chickens. *Avian Pathol.*, 7, 407-409.

Dutta, S. K. and Pomeroy, B. S. (1969). Isolation and characterisation of an enterovirus from baby chicks having an enteric infection. II. Physical and chemical characterisation and ultrastructure. *Avian Dis.*, 11, 9-15.

Endo-Munoz L. B. (1990). A western blot to detect antibody to avian reovirus. *Avian Pathol.*, 19, 477-487.

Fahey, J. E. and Crawley, J. F. (1954). Studies on chronic respiratory diseases of chickens. II: Isolation of a virus. *Can. J. Comp. Med.*, 18, 13-21.

Giambrone, J. J. (1985). Vaccinating pullets to control reovirus associated diseases. *Poult. Digest*, 44 (517), 96-100.

Giambrone, J. J. and Hathcock, T. L. (1991). Efficacy of coarse-spray administration of a reovirus vaccine in young chicks. *Avian Dis.*, 35, 204-209.

Goodwin, M. A., Davis, J. F., McNulty, M. S., Brown, J. and Player, E. C. (1993). Enteritis (so-called runting stunting syndrome) in Georgia broiler chicks. *Avian Dis.*, 37, 451-458.

Jones, R. C. and Onunkwo, O. (1978). Studies on experimental tenosynovitis in light hybrid chickens. *Avian Pathol*, 7, 171-181.

Jones, R. C. and Georgiou, K. (1984). Reovims-induced tenosynovitis in chickens: the influence of age at infection. *Avian Pathol.*, 13, 441-457.

Jones, R. C. and Guneratne, J. R. M. (1984). The pathogenicity of some avian reoviruses with particular reference to tenosynovitis. *Avian Pathol.*, 13, 173-189.

Jones, R. C. and Georgiou, K. (1985). The temporal distribution of an arthrotropic avian reovirus in the leg of the chicken after oral infection. *Avian Pathol.*, 14, 75-85.

Jones, R. C., Islam, M. R. and Kelly, D. F. (1989). Early pathogenesis of experimental reovirus infection in chickens. *Avian Pathol.*, 18, 239-253.

Kaleta E.F. and Heffels-Redmann U. (eds) (1996). International Symposium on adenovirus and reovirus infections of poultry, 24-27 June, Rauischholzhousen, Germany. Institut für Geflügelkrankheiten, University of Giessen, Germany, 343 pp.

Kerr, K. M. and Olson, N. O. (1969). Pathology of chickens experimentally inoculated or contact-infected with an arthritis-producing virus. *Avian Dis.*, 13, 729-745.

Kibenge, F. S. B. and Wilcox, G. E. (1983). Tenosynovitis in chickens. *Vet. Bull*, 31, 39-42.

Kibenge, F. S. B., Gwaze, G. E., Jones, R. C., Chapman, A. F. and Savage C.E. (1985). Experimental reovirus infection in chickens: observations on early viraemia and virus distribution in bone marrow, liver and enteric tissues. *Avian Pathol.*, 14, 87-98.

Krauss, H. and Ueberschar, S. (1966). Zur Structur eines neuen Geflügel-Orphanvirus. Zentralbl. *Veterinärmed.*, 13, 239-249.

Lee, L. H., Shien, J. H. and Shieh, H. K. (1998). Detection of avian reovirus RNA and comparison of a portion of genome segment S3 by polymerase chain reaction and restriction enzyme fragment length polymorphism. *Res. Vet. Sci.*, 65, 11-15.

McNeilly, F., Smyth, J. A., Adair, B. M. and McNulty M.S. (1995). Synergism between chicken anaemia virus (CAV) and avian reovirus. *Avian Dis.*, 39, 532-537.

McNulty, M. S. (1 9 9 3). Reovirus. In Vims infections in birds (J.B. MeFerran and M.S. McNulty, eds). Elsevier Science Publishers BV, Amsterdam, 181-193.

Marquardt, J., Herman s, W., Schulz, L. C. and Leibold, W. (1983). A persistent reovirus infection of chickens as a possible model of human rheumatoid arthritis (RA). Zentralbl. *Veterinärmed.*, 30, 274-282.

Meanger, J., Wickramasinghe, R., Enriquez, C. E. and Wilcox, G. E. (1999). Tissue tropism of avian reovirus is genetically determined. *Vet. Res.*, 30, 523-529.

Menendez, N. A., Calnek, B. W. and Cowen, B. S. (1975). Experimental egg-transmission of avian reovirus. *Avian Dis.*, 19, 104-111.

Meulemanns, G. and Halen, P. (1982). Efficacy of some disinfectants against infectious bursal disease virus and avian reovirus. *Vet. Rec.*, 111, 412-413.

Montgomery, R. D., Villegas, P. and Kleven, S. H. (1986). Role of route of exposure, age, sex and the type of chicken on the pathogenicity of avian reovirus strain 81-176. *Avian Dis.*, 30, 460-467.

Moradian, A., Thorsen, J. and Julian, R. J. (1985). Single and combined infection of specific-pathogen-free chickens with infectious bursal disease virus and an intestinal isolate of reovirus. *Avian Dis.*, 34, 63-72.

Mukiibi-Muka, G. (1997). Studies on local and systemic antibody responses in chickens to avian reovirus infections. *PhD Thesis*, University of Liverpool, United Kingdom, 278 pp.

Page, R. K., Fletcher, O. J. and Villegas, P. (1982). Infectious synovitis in young turkeys. *Avian Dis.*, 26, 924-927.

Petek, M., Feiluga, B., Borghi, G. and Baroni A. (1967). The crawley agent: an avian reovirus. *Arch. ges. Virusforsch.*, 21, 413-424.

Rau, W. E., Van der Heide, L., Kalbac, M. and Girschick, T. (1980). Onset of progeny immunity against viral arthritis/tenosynovitis after experimental vaccination of parent breeder chickens and cross-immunity against six reovirus isolates. *Avian Dis.*, 24, 648-657.

Roessler, D. E. and Rosenberger, J. K. (1989). *In vitro* and in vivo characterisation of avian reovirus. III. Host factors affecting virulence and persistence. *Avian Dis.*, 33, 555-565.

Schwartz, L. D., Gentry, R. F., Rothenbacker, H. and Van der Heide, L. (1976). Infectious tenosynovitis in White Leghorn chickens. *Avian Dis.*, 20, 769-773.

Van der Heide, L., Kalbac, M. and Hall, W.C. (1976). Infectious tenosynovitis (viral arthritis): influence of maternal antibodies in the development of tenosynovitis lesions after experimental infection of day-old chicks with tenosynovitis virus. *Avian Dis.*, 20, 641-648.

Varela, R. and Benavente, J. (1994). Protein coding of assignment of avian reovirus strain S1133. *J. Virol.*, 70, 2974-2981.

Xie, Z. X., Fadl, A. A., Girschick, T. and Khan, M. I. (1997). Amplification of avian reovirus RNA using the reverse transcriptase-polymerase chain reaction. *Avian Dis.*, 41, 654-660.

Yin, H. S. and Lee, L. H. (1998). Development and characterisation of a nucleic acid probe for avian reoviruses. *Avian Pathol.*, 27, 423-426.

Chapter 2

Newcastle Disease/Ranikhet Disease

Newcastle disease is the most dreadful infectious disease of poultry. It is an acute, mild to severe, highly infectious and pathogenic disease of birds caused by avian paramyxovirus -1. It is distributed worldwide and has potential to cause heavy economic losses in the poultry industry (Lancaster, 1976; Spradbrow, 1988). The virus is able to infect over 240 species of birds and it spreads primarily through direct contact between infected and healthy birds (Kaleta and Baldauf, 1988). The disease was first identified at Newcastle-upon-Tyne, England in 1926 and at the same time in Java and Indonesia (Doyle, 1927; Kraneveld, 1926). In India, it was reported from Ranikhet (Uttarakhand) in 1928, hence also named as Ranikhet disease in India.

Etiology

All the available strains of Newcastle disease virus (NDV) belong to the order Mononegavirales, family Paramyxoviridae, and genus Avulavirus. It can be confined to one serotype and are also known as avian paramyxovirus serotype-1. The ND virus genome encodes six structural proteins which are nucleocapsid (N), phosphoprotein (P), matrix (M), fusion (F), haemagglutinin-neuraminidase (HN) and RNA dependent RNA (large) polymerase (L). Due to edition of phosphoprotein (P), it results in another protein named as V protein, which has anti-interferon properties (Czegledi *et al.,* 2006). The ND virus is grouped under avian paramyxovirus type–1 of family Paramyxoviridae. Virus has t-frames/open reading frames (ORFs) which encode the Nucleoprotein (NP), the Phasphoprotein (P), the Matrix protein (M), the fusion protein (F), the haemagglutinine-neuraminidase (HN) and the large protein (L).

On the basis of virulence, there are 3 pathotypes of the virus *viz.* 1. Lentogenic, 2. Mesogenic and 3. Velogenic. Lentogenic pathotype is of low virulence and produces only subclinical infections with mild respiratory and enteric disease symptoms. Mesogenic pathotype is of intermediate virulence which results in respiratory infections with <10 per cent mortality. Velogenic pathotype is highly virulent resulting 100 per cent mortality. It is also called as vvND (very virulent Newcastle disease) which is in a "list A" virus and requires reporting to the Office of International Epizootics. All the three pathotypes are antigenically almost same. Velogenic pathotype is further classified as: 1. Viscerotropic velogenic strain, 2. Neurotropic velogenic strain and 3. Pneumotropic velogenic strain. Viscerotropic velogenic strain causes lethal hemorrhagic lesions in viscera while neurotropic velogenic strain results in encephalitis and pneumotropic velogenic strain results in pneumonic lesions (Lancaster, 1976; Alexander, 1997).

Virus is readily destroyed by 1:5000 $KMnO_4$, 10 per cent formalin and 1:5 ethanol solutions. ND virus (Paramyxovirus) is able to infect and produce diseases in pigeons.

Hosts

ND virus more virulently infects to poultry but less virulently to ducks, turkeys, pheasants and pigeons. According to scientific report NDV is known to infect over 236 species of birds (Kaleta and Baldauf, 1988) and besides poultry species, virulent NDV (vNDV) strains are commonly found in pigeons and double crested cormorants (Diel *et al.*, 2012b; Kim *et al.*, 2008; Pchelkina *et al.*, 2013) and occasionally in some other wild bird species (Kaleta and Kummerfeld, 2012). Typically, the pigeons transmit their vNDV strains of genotype VIb to poultry (Abolnik *et al.*, 2004; Alexander and Parson, 1986). However, poultry are also able to transmit their vNDV strains to pigeons, as well (Merino *et al.*, 2009).

Epidemiology

ND virus is readily excreted in feces, nasal secretions and genital secretions of infected birds. Virus reaches in all the body tissues with in 48-72 hours post infection. Virus is transmitted through feed, water and air. The virus also affects human beings and produces conjunctivitis, headache and flu like symptoms. The virus survives for 12-70 hours at 45°C temperature and for 6 days at 37°C temperature. The lentogenic pathotype ND virus causes 5-50 per cent mortality, mesogenic pathotype ND virus causes 50-55 per cent mortality while velogenic pathotype ND virus causes 90-100 per cent mortality in poultry.

Pathogenesis

The ND virus infection is initiated by its attachment at the surface of the target cells. Binding of the viral HN glycoprotein to the sialic acid containing cell surface proteins, which serves as receptors and triggers the F protein promoted fusion of the viral envelop with the plasma membrane of the host cells through a pH independent mechanism (Lamb and Parks, 2007). After entry, the viral nucleocapsid dissociates the M protein and it is released in to the cytoplasm of the cells going to be infected. Subsequently, the polymerase complex transcribes the viral genomic RNA to produce mRNAs which are required for the synthesis of viral proteins. Binding of polymerase complex to the nucleocapsid is mediated through P proteins, where as the catalytic activities are the function of the L protein (Curran, 1996; Curran *et al.*, 1992; Horikami *et al.*, 1992; Poch *et al.*, 1990; Sidhu *et al.*, 1993).

The switch from transcription to genome replication takes place when sufficient amount of viral proteins have accumulated. The polymerase complex is responsible for the synthesis of full length positive strand antigenomic RNA, which in turn serves as the template for synthesis of negative strand genomic RNA. Viral nucleocapsid is then assembled by association of NP with the newly formed genomic RNA and with the polymerase complex. All complexes of viral proteins are transported to the plasma membrane where they are assembled under the direction of M protein. Viruses are released from the cell by the process of budding (Harrison *et al.*, 2010). Finally, the neuraminidase activity of the HN protein facilitates the detachment of viruses from the infected cells and removes sialic acid residues from progeny virus particles to prevent self-aggregation (Lamb and Parks, 2007; Takimoto and Portner, 2004).

Signs and Symptoms

In VELOGENIC form of disease, per acute mortality is seen without symptoms. Prostration, closed eyes, dropped wings and loss of appetite are some general symptoms in this form. In MESOGENIC form of disease, greenish or yellowish diarrhoea, neck twitching, body temperature is raised to 42°C, comes down to normal and later subnormal range is recorded. Decreased egg production, soft shelled eggs or shell less eggs are laid down by the affected layer birds. Respiratory distress is seen in birds. This form of disease is referred as "Pneumo-encephalitis" form. In LENTOGENIC/MILD form of disease decreased egg production, respiratory distress, gasping and open beaks for breathing through mouth are seen.

Lesions

In VELOGENIC form of disease hemorrhagic ulcers in the ceacal tonsils near ceaco-colon junction are seen. Pinpoint hemorrhages on the tip of proventricular glands are pathognomonic lesions. Hemorrhagic ulcers in intestine, trachea and bronchioles on mucosal surface may be recorded. These hemorrhages are due to endothelial injuries. Congested or ruptured ova may be present in oviducts. In MESOGENIC form of disease, in addition to proventricular glands tip hemorrhages, small and mottled spleen is seen. While in LENTOGENIC form of disease no characteristic lesions are seen.

Diagnosis

Diagnosis is primarily based on signs and symptoms in association with post-mortem lesions. The intestinal villi and crypts undergo coagulative necrosis. Lungs contain serous exudates, get congested and presence of vacuolated macrophages are evident. Liver show proliferation of Kupfer cells, trachea shows mucosal congestion and present of catarrhal exudates. Salpingitis is seen in posterior parts of oviducts. But confirmatory diagnosis is done through isolation of ND virus from tissues of brain, spleen, bone marrow and other involved organs of sick birds. VNT (virus neutralization test), FAT (fluorescence antibody test), CFT (complement fixation test), ELISA and HA/HI (haemagglutination/haemagglutination-inhibition) are serological confirmatory tests. HA/HI tests are adopted for determination of Newcastle disease antibody titer in birds.

Prevention and Control

It is done by adopting better vaccination along with excellent managemental practices. Currently, with modern poultry management systems and use of live attenuated and/or killed/inactivated ND vaccines as per the various programmes and good laboratory support, ND has been effectively controlled. However, due to several epidemiological factors, breaks in biosecurity, inadequate vaccination programmes, immuno-suppressive situations after vvIBD (very virulent infectious bursal disease), and possible transmission of ND from other avian species including pegions, pscittaccines and migratory waterfowls, there are still possibility of ND outbreaks as a re-emerging scenario and so there is a need to take every care for its prevention.

Although vaccination is not a substitute for effective management, there are various national and international policies to control or prevent ND. In some countries, it is a notifiable disease requiring special control measures, enforced by law.

Vaccines and Vaccination

The efficacy of vaccine depends on its invasiveness and power of multiply in the chicken to set up adequate immunity.

A. Mesogenic Live Attenuated Vaccine Strains

1. H-strain: This strain was obtained by passaging a field isolate Hertfordshire (H) through chick embryo and attenuated it to use as vaccine strain. Later, this virulent strain was referred to as Herts'33.

2. Roakin strain: Beaudette (1949) used naturally occurring mesogenic field isolate as mesogenic vaccine.

3. Mukteswar strain: Iyer and Dobson carried out the attenuation of Ranikhet strain in India and further work was continued in India by Hadow and Idnani during 1946. Finally, Dhanda in 1954 developed "Mukteswar strain" which is widely used in India and other Asian countries for effective control of ND.

4. RDV-M strain: It is a mesogenic strain isolated by S. Sulochana and others in 1982 from Mynha, Kerala, India. It is less pathogenic than other mesogenic strains having good immunogenicity and potential for use as ND vaccine.

5. CDF – 4 strain: This strain has been isolated from swine trachea in India having good potentiality to be developed as a mesogenic vaccine.

6. Komorov strain: This mesogenic strain was developed by Dr. Heifa Komorov in 1946 by serial intracerebral passages of field isolate in ducklings.

B. Lentogenic Live Vaccine Strains

1. HB1 or B1 strain: This lentogenic strain was first described by Hitchner in 1948 and used as vaccine.

2. LaSota strain: This lentogenic virus strain of ND was originally isolated by Beaudette in 1946 and used as vaccine strain.

3. F strain: This lentogenic strain is very closely related to B1 strain but slightly less virulent than B1 strain.

LaSota, B1 and F strains are commonly used for vaccination of young chicks at an early age. In general, LaSota vaccine gives better protection than B1 and also has a greater tendency to spread from birds to birds while mesogenic or Komorov strains are used after 8 weeks of age of birds as booster vaccinations.

C. Avirulent Vaccine Strains

V4, V4 – HR and I-2 strains are avirulent strains and produce no disease in poultry but these strains can be used in vaccines for all ages of poultry. V4 – HR is heat resistant strain. V4 – HR and I-2 strains produce thermostable vaccines against NDV infections. The I-2 ND vaccine has been developed for use in controlling Newcastle disease in village chickens. Many ND vaccines are deteriorated after storage for one or two hours at room temperature. This makes them unsuitable for use in village conditions where the vaccine may need to be transported for hours or in some cases days at ambient temperature. The I-2 ND vaccine is more robust and is known as a thermostable vaccine. Thermostable vaccines still require long-term storage in the refrigerator. However during transportation of the vaccine to the field, the vaccine will not be deteriorated as quickly as the traditional vaccines. Evaporative cooling provided by just wrapping the vaccine in a damp cloth will be adequate for maintaining the viability of the vaccine during transportation to remote villages. However if it is stored in direct sunlight or allowed to reach high temperatures (above 37°C) for more than a few hours, it too will be deteriorated resulting its unsuitability for use as vaccine.

D. Inactivated/Killed Vaccines

Since 1930, inactivated/killed ND vaccines are being used for vaccination against Newcastle disease. These vaccines are inactivated by using BPL (Beta–Propriolactone) or formalin and are absorbed on Aluminium hydroxide gel. Oil-adjuvanted vaccines are prepared by using mineral oils. These oil-adjuvented vaccines enhance the immune response in better way because of very slow release of antigens from its deposition sites. Now-a-days a number of different oil emulsion vaccines are available as ND killed or combination with IB (infectious bronchitis) and/or IBD which are being used widely in poultry husbandry.

Live priming followed by killed booster vaccination strategy is used to maintain high levels of antibodies during egg production and to prevent infection. This also allows the transfer of good level of MAbs to their progeny chicks. The MAbs level in progeny chicks and other layer or breeder flocks is determined in terms of HI (Heamagglutination - Inhibition) titers. The FAO bulletin No. 10 gives the HI titer levels (log2) after standard challenge dose. This is a useful guideline for monitoring the flocks and designing the vaccination programmes.

For monitoring a large number of sera samples, computerized automated ELISA antibody technology is being widely used. The sequential serologic and other data are generated in the form of "antibody flock profile" and used to study the periodical vaccine response and other problems in the flocks.

The HI titer (log2) base is protective against contracting ND infections in following ranges:

1. 2 or $<2^2$/(2–4): 100 per cent mortality
2. 2^2 or $<2^5$/(4–32): 10 per cent mortality
3. 2^4 or $<2^6$/(16–64): 0 per cent mortality
4. 2^6 or $<2^8$/(64–256): No mortality but serious drop in egg production.
5. 2^9 or $<2^{11}$/(512–2048): No mortality but there is less drop in egg production.
6. 2^{12} or $<2^{14}$/(4096–16384): No mortality and no drop in egg production.

The half life ($T_{1/2}$) of MAbs raised against ND is generally considered as 4 days. The HI titer of 1:128 is adequate for protection against ND in broiler flocks.

Objectives of Vaccination

1. To establish sufficient local and systemic immunity to prevent morbidity and mortality associated with ND.
2. To prevent damage to the respiratory tract due to ND infection which often allow secondary colonization of respiratory tract by pathogenic bacteria.
3. To prevent egg production drops in laying hens associated with ND infection.
4. To promote the development of high level of antibody titers in breeding flocks during egg laying period intended for transfer in the progeny chicks as MAbs.

Vaccination of Mycoplasma positive flocks with mesogenic and some lentogenic strain vaccines can be associated with sever post vaccination reactions. So, the use of "F" strain vaccine and not "B1" strain vaccine is recommended in Mycoplasma positive flocks. The post vaccination reactions because of live ND vaccine are characterized by tracheal rales, snicking, conjunctivitis and increased susceptibility to E. coli infections. Clinical signs develop within 3-5 days post vaccination and if uncomplicated or mild, subside in next 3-5 days.

Layer/Breeder Vaccination Programme

Programme – 1

The objective of this programme is to stimulate effective mucosal (local) immunity in hens during the egg production period. Here, the level of MAbs (maternally derived antibodies) transferred to the progeny chicks remains generally low and variable. The first vaccination is to be given at 1-3 weeks of age of birds depending on the need of early protection. It primes the bird's immune system to respond maximally for subsequent ND vaccinations. Live ND vaccines induce a local immune response in mucosal associated lymphoid tissues (MALT) producing upper respiratory tract immunity. These vaccines also stimulate systemic immune responses.

The recommended interval between 1st and 2nd vaccinations is 2– 3 weeks because of the short duration of immunity of the primary immune response. Secondary immune responses produced from subsequent vaccinations have a longer duration of immunity. This allows for longer intervals *viz.* 4–6 weeks between booster vaccinations. Flocks previously immunized can be vaccinated with stronger vaccine strains later on. In areas of high challenge risk of velogenic ND, the LaSota strain vaccine can be used for the 1st vaccination, however it may be associated with increased risk of post vaccination reactions. The intra-oular vaccination route is considered as the best rout of administering ND vaccine in comparison to vaccination through drinking water or coarse spray administration within 3 weeks of age of birds, as intra-ocular administration allows each bird to be vaccinated individually resulting in to raising uniform titer.

Commercial ND vaccines are often combined with infectious bronchitis (IB) vaccines due to similarity in vaccination timings and routes of administration. Laying flocks should be vaccinated with live attenuated B1 Newcastle disease vaccine at 30–90 day intervals *i.e.* booster vaccinations to maintain high level of immunity during egg production period and also reducing post vaccination reactions. MAbs can interfere with successful systemic immunization of birds below 3 weeks of age. This interference, however doesn't significantly affect the development of local immunity in the respiratory tract. Single fractioned (monovalent) vaccines of each ND and IB (infectious bronchitis) should not be mixed for vaccination as both the viruses can interfere the replication/multiplication of each other. Only ND and IB combined tailor maid vaccine formulations should be used for the purpose. The effective doses of vaccine strains of ND and IB virus must be properly adjusted in these products to avoid replication/multiplication interference with each other. This programme of using only live attenuated ND vaccines results in good local immunity in respiratory tract and improved immunity in older laying flocks. However, the disadvantages of this programme are requirement of live vaccination of laying birds, chances of post vaccination reactions which may result in to poor egg qualities and reduced egg production with lower level of immunity against ND during peak egg production period.

Programme – 2

The objective of this programme is the use of live attenuated and killed/inactivated ND vaccines to protect the birds during egg laying period in layers and produce high level of uniform MAbs level in the progeny chicks from breeders.

An alternative to the use of only live attenuated vaccines in the laying flocks is to use injectable killed/inactivated ND vaccine prior to the onset of egg production. Birds successfully primed with live attenuated ND vaccines respond better to killed/

inactivated ND vaccination. In laying hens, the killed/inactivated vaccine provides an immunity of long duration.

A minimum interval of 4 weeks between the last live attenuated vaccination and 1st killed/inactivated vaccination is required for an optimum immune response. Killed/inactivated ND vaccines are generally recommended 4 weeks prior to the onset of egg production (*i.e.* at 12th week of age of birds) to allow sufficient time for the development of full immune response. In most situations, killed/inactivated ND vaccines provide immunity sufficient for one laying cycle. For hens being molted, a second booster dose of killed/inactivated vaccine should be given during molting period. However, the residue tissue reactions from vaccination with oil- adjuvanted killed/inactivated vaccines may be detected for many weeks post vaccination. If poor priming of pullets has been done, inadequate or variable immune response will be resulted after killed/inactivated ND booster vaccination. The advantages of this vaccination programme are long duration immunity, sufficient level of immunity during peak egg production period, delivery of many antigens through polyvalent vaccines in a single vaccination and each bird immunization. However, the disadvantages are increased immunization cost and requirement of individual bird handling which may induce stress in the birds.

Programme – 3

The objective of this programme is to immunize the layer and breeder flocks in high risk area of velogenic ND virus challenge.

In this programme, simultaneous live and killed/inactivated ND vaccines are being used which is all most effective against high challenge of velogenic ND virus strains. These vaccines (live attenuated and killed/inactivated) are used simultaneously at an age of 3 weeks, but can be administered as early as 1 day. Live attenuated ND vaccine is used intra-ocularly and killed/inactivated ND vaccine is used in the same pullet simultaneously through injection. This programme is only adopted in high risk areas of velogenic ND virus challenge. Only tailor maid vaccine is recommended, if vaccination is being practiced in the birds of less than 10 days of age. If this ND vaccine is used in hatchery, separate site of injection should be chosen from the site at which Marek's disease vaccine has to be given.

Broiler Vaccination

Programme – 1

The objective of this programme is to protect the broiler flocks in which chicks are having low MAbs level using live attenuated ND vaccines.

Live attenuated ND vaccination at 1 day of age stimulates mucosal immunity in the upper respiratory tract of broilers. The immunity because of the hatchery vaccination is of short period *viz.* 3 – 5 weeks and only limited to the upper respiratory tract. Revaccination (2nd live vaccination) is generally necessary to provide sufficient immunity throughout the growing period. A 3rd live vaccination in broilers to be kept for longer period of time or in birds of area at risk of high challenge is recommended. Which types of vaccine strains are required, it depends upon the vaccination programme to be adopted (*i.e.* higher immunogenicity or less reactivity).

Programme – 2

The objective of this programme is to vaccinate the broiler flocks in which chicks are having moderate to high MAbs levels using live attenuated ND vaccines.

The advantage of vaccinating older broiler birds are the MAbs interference is less likely to occur and the bird's immune system is more developed. Early protection against infections is provided by MAbs. The 1st vaccination usually utilizes milder vaccines applied by a milder method of administration. But LaSota strain containing vaccine may be used in the areas of high challenges. Revaccination is usually necessary to provide sufficient immunity throughout the growing period. A 3rd live vaccination may be needed in longer lived broilers (roasters) or in birds of area at risk of high challenge.

Programme – 3

The objective of this programme is to protect the broiler flocks using live and killed/inactivated ND vaccines simultaneously in the area of high risk of velogenic ND virus challenge.

The live attenuated ND vaccines usually contain milder virus strains and administered intra-ocularly. Simultaneously killed/inactivated ND vaccine appropriately tailor-made for young birds is used. However, re-vaccination depending upon the current HI titer of the flocks and time period for which the flocks have to be kept is usually considered in high very virulent ND (vvND) virus challenge areas. It also decides the use of live attenuated or killed/inactivated ND vaccines depending upon the need.

REFERENCES

Abolnik, C., Horner, R. F., Maharaj, R. and Viljoen, G. J. (2004). Characterization of pigeon paramyxovirus (PPMV-1) isolated from chickens in South Africa. *Onderstepoort J. Vet. Res.*, 71, 157-160.

Alexander, D. J. (1997).New castle disease and other avian paramyxoviridae infections. In *Disease of Poultry*, 10th edition. Edited by: Calnek, B. W., Barnes, H. J., Beard, C. W., McDougald, L. R., Saif, Y. M., Ames, Iowa: Iowa State University Press, 541 – 570.

Alexander, D. J. and Parson, G. (1986). Pathogenecity for chickens of avian paramyxovirus type-1 isolates from pigeons in Great Britain during 1983-1985. *Avian Pathol.*, 15, 487-493.

Curran, J. (1996). Re-examination of the Sendai virus P protein domains required for RNA synthesis: a possible supplemental role for the P protein. *Virology*, 221, 130 - 140.

Curran, J., Marq, J. B. and Kolakofsky, D. (1992). The Sendai virus nonstuctural C proteins specifically inhibit viral mRNA synthesis. *Virology*, 189, 647 – 656.

Czegledi, A., Ujvari, D., Somogyi, E., Wehmann, E., Warner, O. and Lomniczi, B. (2006).Third genom size categoryof avian paramyxovirus serotype 1 (Newcastle disease virus) and evolutionary implications. *Virus Res.*, 120, 36 – 48.

Diel, D. G., Miller, P. J., Wolf, P. C., Mickley, R. M., Musonte, A. R., Emanueli, D. C., Shively, K. J., Pedersen, K. and Afonso, C. L. (2012b). Characterization of Newcastle disease virus isolated from cormorant and gull species in the United states in 2010. *Avian. Dis.*, 56, 128-133.

Doyle, T. M. (1927). A hitherto unrecorded disease of fowls due to a filter-passing virus. *J. Comp. Pathol.*, 40, 144 – 169.

Harrison, M. S., Sakaguchi, T. and Schmitt, A. P. (2010). Paramyxovirus assembly and budding: building particles thattransmit infections. *Int. J. Biochem. Cell Biol.*, 42, 1416 – 1429.

Horikami, S. M., Curran, J., Kolakofsky, D. and Moyer, S. A. (1992). Complexes of Sendai virus NP-P and P-L proteins are required for defective interfering particle genome replication *in vitro. J. Virol.*, 66, 4901 – 4908.

Katela, E. F. and Baldauf, C. (1988). New castle disease in free-living and pet birds. In *Newcastle Disease*. Edited by Alexander, D. J., Boston: Kluwer Academic Publishers, 197 – 246.

Kaleta, E. F. and Kummerfeld, N. (2012). Isolation of Herpesvirus and Newcastle Disease virus from White Storks (Ciconia ciconia) maintained at four rehabilitation centres in northern Germany during 1983 to 2001 and failure to detect antibodies against avian Influenza A virus of serotype H5 and N7 in the birds. *Avian Pathol.*, 41, 383-389.

Kim, L. M., King, D. J., Guzman, H., Tesh, R. B., Travassos da Rosa, A. P., Bueno, Jr. R., Dennett, J. A. and Afonso, C. L. (2008). Biological and phylogenetic characterization of pigeon paramyxovirus serotype-1 circulating in wild North America pigeons and doves. *J. Clin. Microbiol.*, 46, 3303-3310.

Kraneveld, F. C. (1926). A poultry disease in the Duch East Indies. *Ned-Ind BI Diergeneesk*, 38, 450 – 488.

Lamb, R. A. and Parks, G. D. (2007). Paramyxoviridae: the viruses and their replications. In *Fields Virology*-5th edition. Edited by Knipe, D. M., Howley, P. M., Griffin, D. E., Lamb, R. A., Martin, M. A., Roizman, B., Straus, S. E., Philadelphia, P. A., Lippincott, W., 1449 – 1496.

Lancaster, J. E. (1976). A history of Newcastle disease with comments on its economic effects. *World Poultry Sci. J.*, 32, 167 -175.

Mayo, M. A. (2002). A summary of taxonomic changesrecently approved by ICTV. *Arch. Virol.*, 147, 1655 – 1663.

Merino, R., Villegas, H, Quintan, J. A. and Calderon, N. (2009). Characterization of Newcastle disease viruses isolated from chicken, gamefowl, pigeon and quail in Mexico. *Vet. Res. Commun.*, 33, 1023-1030.

Pchelkina, L. P., Manin, T. B., Kolosov, S. N., Starov, S. K., Andriyasov, A. V., Chvala, I. A., Drygin, V. V., Yu, Q., Miller, P. J. and Suarez, D. L. (2013). Characterization of pigeon paramyxovirus serotype-1 isolates (PPMV-1) from the Russian Federation from 2001 to 2009. *Avian Dis.,* 57, 2-7.

Poch, O., Blumberg, B. M., Bougueieret, L. and Tordo, N. (1990). Sequence comparison of five polymerases (L proteins) of unsegmented negative-strained RNA viruses: theoretical assignment of functional dmains. *J. Gen. Virol.,* 71, 1153 – 1162.

Sidhu, M. S., Menonna, J. P., Cook, S. D., Doeling, P. C. and Udem, S. A. (1993). Canine distemper virus L gene: sequence and comparison with related viruses. *Virology,* 193, 50 – 65.

Spradbrow, P. B. (1988). Geographical distribution. In: *Newcastle Disease.* Edited by: Alexander, D. J., Boston: Kluwer Academic Publishers, 247 – 255.

Takimoto, T. and Portner, A. (2004). A molecular mechanism of paramyxovirus budding. *Virus Res.,* 106, 133 – 145.

Chapter 3

Marek's Disease

Marek's disease (MD) is a contagious viral disease of domestic fowl (chickens) caused by an Alphaherpesvirus (Davison and Nair eds., 2004, Schat and Nair, 2008, Sharma, 1998).This virus is also called as Gallid Herpesvirus–2 (GaHV–2) belongs to the genusMardivirus of family Herpesviridae and subfamily Alphaherpesvirinae. Marek's disease virus (MDV) was earlier classified with in the subfamily Gammaherpesviridae because of its biological properties but was reclassified in 2002 after complete sequencing of its genome in the new genus Mardivirus (Marek's disease like virus) for which it became the type species (Tulman et al., 2000). MD is reported to occur worldwide. But this is not a notifiable disease (as in Newcastle disease). MD results in suntantial economic losses estimated as more than 1 billion $ per year (Morrow and Fehler, 2004). Although MD was described in 1907 by Joseph Marek, the MDV was only isolated in 1967 in the United Kingdom (Churchill and Biggs, 1967). However, the 1st report of MD in United Kingdom was made in a meeting at Royal society of medicine by Galloway in 1929.

MDV infectious particles comprise more than 30 different proteins assembled according to a complex architecture including the following: (1). A central capsid containing the viral genome, (2). A protein layer termed "tegumen" comprising >15 proteins and (3). A lipid bilayer in which about 10 enveloped glycoproteins are anchored. The MDV genome is a linear double–strainded dsDNA of approximately 175 kbp which contains a uniquely long (UL) sequence and a unique short (US) sequence (Tulman et al., 2000).

MDV replicates efficiently only in primary chicken or duck cells in culture (Chuchill and Biggs, 1967, Nazerian et al., 1968), yielding titers between 10^5 and 10^7 p.f.u./ml depending upon the strain. MDV infections are performed by co-culturing infected cells with naive cells because the virus cannot be purified as cell free virus from cell lysate or culture supernatants. This defferent characteristic constitutes constraints for vaccine production.

Epidemiology

Birds got infected by inhalation of contaminated dust from the poultry houses and following a complex life-cycle, the virus is shed from the feather follicles of infected birds (Baigent and Davison, 2004). MD occurs at 3–4 weeks of age or older and is most common between 12– 30 weeks of age. Congenital infection doesnot occur and chicks are protected by MAbs for first few weeks of life (Frederick *et al.*, 1999, Clarence *et al.*, 1986, Quinn *et al.*, 2002). Epidemics involve sexually immature birds of 2 to 5 months of age, high mortality rate about 80 per cent soon peaks and then decline (Frederick *et al.*, 1999). However, the mortality rate may vary from 1 per cent -50 per cent during the lifespan of chicken in the population (Purchase *et al.*, 1976).

In the period of 2–7 days post-infection, MDV multiplies in B-lymphocytes resulting in B-cell death. This activates T-lymphocytes. At the same time, antibody production decreases. Then the infection moves to the activated T-cells and the cellular immunity is compromised. From 6-14 days, the immune response develops and the virus then goes to hiding (latency) and at 14 day, the virus appears in the feather follicles and spread to the environment to infect new hosts. After 14 days, the virus may emerge again in the original host and replicates. MDV may transform lymphocytes, form tumors and cause death of the hosts.

Pathogenesis

MDV enters chicken via the respiratory tract after inhalation of contaminated dust and dander originated from infection harbouring poultry houses (Baigent, *et al.*, 2013). Then the virus infects B-lymphocytes and macrophages in the lungs (Baaten *et al.*, 2009; Calnek, 2001) and then transported towards the main lymphoid organs *viz.* bursa of Fabricius, thymus and spleen. The primary target cells in al these organs are bursa-derived lymphocytes (B-cells), although some thymus-derived lymphocytes (T-cells) firstly helper T-cells (CD4+) followed by cytotoxic T-cells (CD8+) may also be infected leading to compromised humoral and cellular immunity resulting in ti immuno-suppression. Now immuno-suppressed birds are susceptible to any type of infections perticularly secondary bacterial infections. After replicating in B-lymphocytes, MDV infects activated T-lymphocytes mainly CD4+ cells. It is believed that only a few T-lymphocytes undergo transformation and becomes the origin of T-lymphomas which may be either monoclonal or oligoclonal (Mwangi *et al.*, 2011). The lymphomas are mainly localized in visceral organs like kidneys, spleen, liver, gonads, proventriculus, peripheral nerves, skin and muscles. In most transformed T-lymphocytes, MDV remains in latent phase and doesnot replicates to produce virus particles. Only a small portion of tumor cells (<0.01 per cent) express lytic viral antigens and contains viral particles. MDV undergoes for latency only in lymphocytes and not in neurons. During early infection, the virus is transported towards the skin, most specifically to feather follicles (Baigent, *et al.*, 2005). From these infected feather follicles, MDV sheds in to the environment via scales, feather debris and danders, which becomes the major source of infection to other birds. Bird to bird transmission is exclussively horizontal. There is no vertical transmission from hens to the progeny chicks via eggs.

Signs, Symptoms ans Lesions

There are four forms of disease depending upon the involvement of organs in infection:

1. Cutaneous/Skin form
2. Classical/Neural form
3. Acute/Visceral form
4. Ocular/Gray eye form

1. Cutaneous/Skin Form

Before understanding of this form of MD, one should go for the study of skin as how the interrection between skin cells and MDV is established? Bird's skin differs from that of mammalsby its thinness, by the presence of feathers instead of hairs and by the absence of sebaceous glands, although the overall histological structure is similar (Lucas and Stettenheim, 1972; Spearman and Hardy, 1985). Bird's skin is composed of an epidermis separated from dermis by a basal membrane. This basal membrane is a thin and continuous layer which serves as a molecular filter and anchoring point for the epidermis basal cells via hemidesmosomes. Bird's dermis is relatively thin in comparison with mammals. The epidermis is a multistratified, keratinized squamous epithelium, whose thickness varies depending upon the region of the body. The deep layer of epidermis *i.e.* stratum germinativum is composed of live cells arranged in three layers *i.e.* the basal, the intermediate and the transitional layers.The basal layer which is just above the basal membrane (below the basal membrane dermis is present) is constituted of small undifferentiated cubic cells having high deviding rate and which migrates towards more superficial layers. The intermediate layer is constituted of cubic cells that have migrated from basal layer. The transitional layer is constituted of 2 or 3 layers of flat elongated cells containing a large number of intracellular lipid vacuoles or droplets which is typical of bird's skin.

The external layer or cornified layer of epidermis (also called stratum corneum) is composed of corneocytes, which are flat, dead, anucleated and keratinized cells organized in sheets. The differentiation of basal cells in corneocytes is a normal physiological process in the epidermis. The main cellular modifications are the loss of organells, formation of lipid vacuoles deposition of keratinized fibers in the cytoplasm and a thick envelop under the plasma membrane (Spearman and Hardy,1985).Corneocytes which are detached regularly from the epidermis, are constantly renewed by the cells from the lower layers. This process is called "exfoliation or desquamation."

As in mammals, chicken epidermis contains dendritic cells (Langerhans cells) whose number is estimated as 8000 per mm[2] of epidermis in an 8 weeks aged chick (Perez and Millan, 1994, Igyarto *et al.*, 2006). Following antigenic stimulation, these dendritic/Langerhans cells seem to migrate to dermal lymphoid nodules and not to lymph nodes as lymphnodes are not present in the birds (Igyarto *et al.*, 2006). Besides feathers, bird's epidermis contains melanocytes including in non-coloured chickens also.

Feather Follicles

Birds are the only animals for which feathers are absolutely necessary not only for flying but also for thermal regulation barrier of body. Feathers are the most complex and most diversified integumentary products found in vertibrates. Feathers are exclusively composed of beta–keratin (Presland *et al.*, 1989) and arise from the feather follicles. The feather follicles are formed by the invagination of the epidermis around the "feather filament cylinder" in to the dermis at day 14 of embryogenesis, which losts 21 days in chicken.

There are as many feather follicles as there are feathers on the skin *i.e.* between 20,000 and 80,000 depending upon the bird's species (Yu *et al.*, 2004). At the base of the feather follicles, three structures namely 1. Dermal papila, 2. Epidermal collar and 3. Collar buldge are present. Follicle stem cells, which are located in the collar buldge, give rise to a population of "Transient Amplifying" (T.A.) cells, which allow the renewal of feather and the follicle after molting or after accidentally plucking of feathers (Yue *et al.*, 2005; Lin *et al.*, 2006). Repeated molting ensures the regular renewal of bird feathers throughout its life span. Feather follicles contain melanocytes responsible for the colour of the feather.

Horizontal Transmission of MDV

Under natural conditions, MD transmission is air borne (Sevoian *et al.*, 1963; Biggs and Payne, 1967) suggesting that the virus is excreted and relatively resistant in the external environment. The presence of infectious virions in the skin and feather follicles of infected chickens was confermed (Calnek *et al.*, 1970; Nazerian and Witter, 1970). Feather follicle can produce complete mature infectious virions, harbouring a tegument and an envelope. The infectiousness of MDV in the environment can lost up to 7 months at room temperature (Carrozza *et al.*, 1973) and 16 weeks in the litters (Witter *et al.*, 1968). The above facts suggest that MDV is transmitted from birds to birds through direct or indirect contact with feather tips if lost or plucked from the body. This is called as "horizontal transmission" of MDV.

The epithelium of feather follicles is the tissue where the viral antigen expression is the most commonlly and positively found than other tissues (Calnek and Hitchner, 1969; Purchase, 1970). This tissue expresses the highest level of viral antigens for the longest period of time. These antigens are located in the upper layers of stratum germinativum of feather follicles. Viral antigens are detectable in feather follicles from feather's tips between 11 to 14 days post-infection using standard biochemical methods (Malkinson *et al.*, 1989; Nilkura *et al.*, 1999).

Methods of MDV Detection in Skin or Feathers

Viral antigens are detected by immuno-fluorescence on tissue sections (Calnek and Hitchner, 1969) or by gel immuno-deffusion or E.L.I.S.A. from feather tip cell extracts (Purchase, 1970; Davidson *et al.*, 1986). Polymerase chain reaction (P.C.R.) (Davidson and Borenshtain, 2002), pulse field gel electrophoresis (P.F.G.E.) (Davidson and Borenshtain, 2003), loop-mediated isothermal amplification (L.A.M.P.) test (Wozniakowski *et al.*, 2011) are other usefull tests for MDV detection from feather follicles. A virus neutralization test (V.N.T.) is also used for test the ability of serum to neutralize the plaque forming property of cell free MDV. ELISA is also readily available to detect antibodies generated against MDV (Zelnik *et al.,* 2004). However, the mechanism by which MDV infects skin and feather follicles are poorly understood. Because B and T lymphocytes are the major targets of MDV and are infected easily, it seems that probably these cells are the vehicle for feather follicles infection (Calnek, 1986; Calnek, 2009). For the most pathogenic strains, replication of MDV starts at 1 week post-infection in the feather follicles, well before tumor development.

Two types of lesions are found in skin: 1. Tumor like lesions and 2. Non-tumor like lesions. In tumor like lesions, birds presented hypertrophoid feather follicles with compact lymphoid aggrigates in the dermis associated with cappilaries upon microscopic examination. Among non-tumor like lesions, the nuclear inclusion bodies are typically found during herpesvirus infections. These nuclear inclusion bodies are only found in the upper layers of the feather follicle epithilium and never in the basal layer (Lapen *et al.,* 1970). Cutaneous disease involves round, nodular lesions up to 1 cm in diameter particularly at feather follicles (Frederick *et al.,* 1999).

2. Classical/Neural Form

This form results in tumors in the nerves and central nervous system (C.N.S.) specially brain. Tumors of the nerves will cause weakness to paralysis of wings and legs causing the chickens to have trouble in moving and thereby unable to get feed and water. Ultimately, the birds affected die due to dehydration and starvation. This form is also reffered as classical form of MD. Mortality rarely exceeds 10–15 per cent.

The characteristic finding is enlargement of one or more peripheral nerves. Those most commonly affected and easily seen at post-mortem are the brachial plexuse, sciatic plexuse, coeliac plexuse, abdomenal, vagus and intercostal nerves. Affected nerves are often 2– 3 times their normal thickness, the normal cross-striated and glistening appearance is absent and the nerve may appear grayish or yellowish and sometimes oedomatous. Lymphomas are sometimes present also in the classical form of MD, most frequently as small and soft gray tumors in the ovary and sometimes in the lungs, kidneys, heart liver and other tissues. Perivascular cuffing of lymphoid cells of various thickness are found in central nervous system (C.N.S.) by very virulent MDV (vvMDV) resulting in to non-suppurative meningo-encephalomyelitis and lymphoma lesions in brain (Cho *et al.,* 1998).

3. Acute/Visceral Form

This form of MD causes tumors of the internal organs such as liver, spleen, gonads, feather follicles (skin form), heart, kidney and proventriculus. This form is also termed as "acute form". Here the typical finding is widespread diffused lymphomatous involvement of the liver, gonads, spleen kidneys, lungs, proventriculus and heart. Sometimes lymphomas also arise in the skin around the feather follicles and in the skeletal muscles. In younger birds, liver enlargement is usually moderate but in adult birds, the liver may be greatly enlarged like a disease Avian Leukosis Complex (A.L.C.), but both the diseases are differentiated microscopically as described in the table given below. Nerve lesions are often absent in adult birds.

In both classical and acute forms of MD, the disease starts as proliferation of lymphoid cells, which is progressive in some cases and retogressive in others. The peripheral nerves may be affected by proliferative, inflammatory or minor infiltrative changes which are termed type A, B and C lesions, respectively. The "A" type lesions consist of infiltration by proliferatinf lymphoblasts, large, medium and small lymphocytes and macrophages which appear as neoplastic in nature. The "B" type lesion is characterized by interneuritic oedema, infiltration mainly by small lymphocytes and plasma cells, Schwann cells proliferation and appears to be inflammatory in nature. The type "C" lesion consists of a light scattering of mainly small lymphocytes and is often seen in birds that show no gross lesions or clinical signs. So, the type "C" infection is thought to be as a retrogressive inflammatory lesion. Demyelination frequently occurs in nerves affected by type "A" and "B" lesions, which is responsible for clinical paralysis.

4. Occular Form

In this form of MD, "grey eye" is caused by an iridocyclitis that renders the bird unable to accommodate the iris in response to light and cause a distorted pupil. It is common in older (16 – 18 weeks of age) birds and it may be the only presenting sign. The colour of iris is changed from normal orange/red/yellow to grey. It may be resulte in to irregularly shaped pupil.

Treatment

No specific treatment is available.

Prevention and Control

The disease is prevented up to a better extent by using proper vaccines in a proper way and at proper time. However, the control needs better managemental systems at farm. Some managemental aspects to be followed are:

1. Isolate affected bird(s) to an area away from ther healthy chickens.
2. Visit the isolated affected chickens in the last at the end of day.
3. Call for a local veterinarian.
4. Donot put young and unvaccinated birds in a pen from which affected birds are housed.
5. Before introducing a new flock of chickens, remove dust, feathers, fecal material andlitter from the infected pen. Clean the house thoroughly and follow up by sanitizing with 10 per cent Chlorine solution on all cleaned non-porous surfaces such as roosts, wire enclosures, nets, feed pans and watcrers.
6. Effective biosecurity is applied in the farm.
7. No visitor should be allowed to enter the farm if he is coming form an affected farm(s).
8. Each bird should be effectively vaccinated.

Differentiation between MD and ALC

Sl.No.	Feature	MD	ALC
1.	Age	Any age, usually 6 weeks or older	Not under 16 weeks
2.	Signs	Frequently paralysis	Non-specific
3.	Incidence	Frequently above 5 per cent in non-vaccinated flocks. Rare in vaccinated flocks.	Rarely above 5 per cent
4.	Neural involvement	Frequent	Absent
5.	Bura of Fabicius	Interfollicular tumors	Intrafollicular tumors
6.	Tumors of skin, muscle and proventriculus	May be present	Usually absent
7.	Grey eye	May be present	Usually absent
8.	Liver tumors	Often perivascular	Focal or diffuse
9.	Spleen	Diffuse	Often focal
10.	CNS system involvement	Yes	No
11.	Lymphoid proliferation in skin and feather follicles	Yes	No
12.	Tumor cytology	Lymphoblasts, large, medium and small lymphocytes	Only lymphoblasts
13.	Category of neoplastic lymphoid cell	T cell	B cell

Vaccines and Vaccination

MD remains an economically important disease in chickens even in the face of successful vaccination programmes. One of the key problems in protection against MD, is that chicks are challlenged with MDV as soon as they are placed on the farms, which is in general within 1 -3 days after hatching. Vaccine – induced protection can be demonstrated as early as 5 days post vaccination (D.P.V.) under controlled conditions, *e.g.* in laboratory challenge experiments (Witter *et al.*, 2001b). Vaccination at embryo day 18th (Sharma and Burmester, 1982) was developed based on the hypothesis that an increase in thr time between vaccination and the earliest possible challenge may allow the development of a more solid protective immunity. Although in ovo vaccination for MD is now widely and successfully used in USA (Avakian *et al.*, 2000).

There are "3" serotypes of MDV based on the antigenically relatedness: 1. Serotype – 1, 2. Serotype – 2. and 3. Serotype – 3.

1. Serotype–1

This includes all the pathogenic strains of MDV. These strains are: very virulent plus (*e.g.* 648A), very virulent (*e.g.* Md/5, Md/11, Ala-8, RB-1B), virulent (*e.g.* HPRS-16, JMGA), mildly virulent (*e.g.* HPRS–B14, Conn A) and finally to weakly virulent (*e.g.* CU-2, CVI-988). These strains may be attenuated by passaging in tissue cultures resulting in to loss of pathogenic properties but retaintion of immunogenicity. Those that are commerciallyused include attenuated HPRS-16 and CVI-988 (Rispens) strains. Serotype–1 vaccines are prepared in a cell associated (wet) form that must be stored in liquid nitrogen at a temperature of -196°C.

2. Serotype–2

This include naturally avirulent strains of MDV (*e.g.* SB-1, HPRS-24, 301B/1, HN-1) and several of these have been shown to provide protection against virulent strains. The SB-1 and 301B/1 strains have been developed commercially and used particularly with HVT (Herpes Virus of Turkey) as bivalent vaccines for protection against the very virulent strains. Serotype 2 vaccines exist only in the cell-associated (wet) form and remains stored in liquid nitrogen at -196°C temperature. These SB1 strain vaccines are not immunogenic alone but when used with serotype 3 *i.e.* HVT strain, acts synergistically to stimulate potential immune responces.

3. Serotype–3

This contains the strains of naturally avirulent HVT (*e.g.* FC 126, PB1), which are widely used as a monovalent vaccine and also in a combination with serotype 1 and 2 strains as bivalent or trivalent vaccines against the very virulent MDV (vvMDV) strains. HVT may be prepared in a cell-free form as a freeze-dried (Lyophilized) vaccine or in a cell-associated (wet) form and stored in liquid nitrogen (LN_2) at –196°C temperature. Handling of this Liquid Nitrogen (LN_2) preserved vaccines should be done only after wearing specific hand gloves.

Commercial vaccines have been developed from all the three serotypes of MD virus. Serotype -1 and serotype -3 vaccines are used as monovalent or combination vaccines. Serotype -2 vaccines are used with serotype -3 vaccines as bivalent vaccines. Due to protective synergism, the combinations of vaccines of serotype-2 and serotype–3 often provide greater protection than monovalent vaccines alone. Serotypes -1, 2 and 3 are given sometimes as trivalent vaccines. These bivalent and trivalent vaccines are used in high challenge areas and also where chances of vvMD.

Serotype	Strain	Species of Origin	Ability to Spread Horizontaly	Used as Monovalent Vaccine
1	CVI-988	Chicken	+	+
1	CVI-988/C	Chicken	+	+
1	CVI-988/CR6	Chicken	+	+
1	MD11/75C	Chicken	-	+
2	SB1	Chicken	+	-
2	301B/1	Chicken	+	-
3	FC126 (HVT)*	Turkey	-	+

*HVT = Herpes Virus of Turkey, Two forms: 1. Cell-free and 2. Cell-associated.

Layers/Breeders/Broilers

Vaccination Programme

The objective of MD vaccination is to prevent the mortality, morbidity and immuno-suppression associated with the neural (classic), visceral (acute), acular and cutaneous forms of MD in commercial layers and breeders. In broilers, to prevent condemnations for viscersl or cutaneous forms of MD at the time of processing and to improve growth performance. This live vaccination programme is for use in commercial egg layers, breeders and broilers when protection from MD is required.

MD vaccines are administered at the hatchery either invivo soon after hatching subcutaneously or inovo on day 18 of incubation of fertilized eggs. Vaccination later than 1 day of age is less efficacious. Revaccination of layer and breeder pullets after 1 day age is practiced in some areas with high challenges of MDV.

Serotype -2 vaccines (SB1 and 301B/1) used alone without HVT (FC126), are not efficacious. Poor vaccination technique leaving birds unvaccinated can result in MD outbreaks. Chicks placed in an environment heavily contaminated with pathogenic MDV may not be protected by vaccination. Vaccine should not be used as diluted which is intended to be used in commercial egg layers and breeders. Overdilution of MD vaccine used in broilers can result in disease outbreaks. Addition of other vaccines, vitamins, dyes or antimicrobials not approved for combination with MD vaccination can decrease the potency of MD vaccines. Single dose MD vaccination is sufficient for life time whether it is cell-free or cell-associated vaccine.

REFERENCES

Avakian, A. P., Wakenell, P. S., Grosse, D., Whitfill, C. E. and Link, D. (2000): Protective immunity to infectious bronchitis in broilers vaccinated against Marek's disease either inovo or at hatch and against infectious bronchitis at hatch. *Avian Dis.*, 44: 536 – 544.

Baaten, B. J., Staines, K. A., Smith, L. P., Skinner, H., Davison, T F. and Butter, C. (2009): Early replication in pulmonary B cells after infection with Marek's disease herpesvirus by the respiratory route. *Virol. Immunol.*, 22: 431 – 444.

Baigent, s. J., Kgosana, L., Gamawa, A., Smith, L. P., Read, A. F. and Nair, V. (2013): Relationship between levels of very virulent MDV in poultry dust and in feather tips from vaccinated chickens. *Avian Dis.*, 57: 440 – 447.

Baigent, S. J., Smith, L. P., Currie, R. J. and Nair, V. K. (2005): Replication kinetics of Marek's disease vaccine virus in feathers and lymphoid tissues using PCR and virus isolation. *J. Gen. Virol.*, 86: 2989 – 2998.

Biggs, P. M. and Payne, L. N. (1967): Studies on Marek's disease. 1. Experimental transmission. *J. Natl. Cancer Inst.*, 39:267 – 280.

Calnek, B. W. (1986): Marek's disease – a model for herpesvirus oncology. *Crit. Rev. Microbiol.*, 12: 293 – 320.

Calnek, B. W. (2001): Pathogenesis of Marek's disease virus infection. *Curr. Top. Microbiol. Immunol.*, 255: 25 – 55.

Calnek, B. W., Adldinger, H. K. and Kahn, D. E. (1970): Feather follicle epithelium: a source of enveloped and infectious cell-free herpesvirus from Marek's disease. *Avian Dis.*, 14: 219 – 233.

Calnek, B. W. and Hitchner, S. B. (1969): Localization of viral antigen in chickens infected with Marek's disease herpesvirus. *J. Natl. Cancer Inst.*, 43: 935 – 949.

Carrozza, J. H., Fredrickson, T. N., Prince, R. P. and Liginbuhl, R. E. (1973): Role of desquamated epithelial cells in transmission of Marek's disease. *Avian Dis.*, 17: 767 – 781.

Churchill, A. E. and Biggs, P. M. (1967): Agent of Marek's disease in tissue culture. *Nature*, 215: 528 –530.

Davidson, I. and Borenshtain, R. (2002): The feather tips of commercial chickens are a favorable source of DNA for the amplification of Marek's disease virus and avian leukosis virus, subgroup. *J. Avian Pathol.*, 31:237 – 240.

Davidson, I. and Borenshtain, R. (2003): Novel applications of feather tip extracts from MDV-infected chickens; diagnosis of commercial broilers, whole genome separation by PFGE and synchronic mucosal infection. *FEMS Immunol. Med. Microbiol.*, 38: 199 – 203.

Davidson, I., Maray, T. Malkinson, M. and Becker, Y. (1986): Detection of Marek's disease virus antigens and DNA in feather from infected chickens. *J. Virol. Methods*, 13: 231 – 244.

Igyarto, B. Z., Lacko, E., Olah, I. and Magyar, A. (2006): Characterization of chicken epidermal dendritic cells. *Immunology*, 119: 278 – 288.

Lapen, R. F., Piper, R. C. and Kenzy, S. G. (1970): Cutaneous changes associated with Marek's disease of chickens. *J. Natl. Cancer Inst.*, 45: 941: 950.

Lin, C. M., Jiang, T. X., Widelitz, R. B. and Chuong, C. M. (2006): Molecular signaling in feather morphogenesis. *Curr. Opin. Cell. Biol.*, 18: 730 – 741.

Lucas, A. M. and Stettenheim, P. R. (1972): Avian Anatomy Integument. *Agriculture Handbook*, 362: 346 – 629.

Malkinson, M., Davidson, I., Strenger, C., Weisman, Y., Maray, T., Levy, H. and Becker, Y. (1989): Kinetics of the appearance of Marek's disease virus DNA and antigens in the feathers of chickens. *Avian Pathol.*, 18: 735 – 744.

Morrow, C. and Fehler, F. (2004): Marek's disease: a worldwide problem. In Marek's disease: an evolving problem. Edited by Davison, F., and Nair, V. London: Elsevier Academic Press, p. 49 – 61.

Mwangi, W. N., Smith, L. P., Baigent, S. J., Beal, R. K., Nair, V. and Smith, A. L. (2011): Clonal structure of rapid-onset MDV-driven CD4+ lymphomas and responding CD8+ T cells. *PloS Pathog.*, 7:e1001337.

Nazerian, K., Solomon, J. J., Witter, R. L. and Burmester, B. R. (1968): Studies on the etiology of Marek's disease. II. Finding of a herpesvirus in cell culture. *Proc. Soc. Exp. Biol. Med.*, 127:177 – 182.

Nazerian, k. and Witter, R. L. (1970): Cell-free transmission and invivo replication of Marek's disease virus. *J. Virol.*, 5: 388 – 397.

Niikura, M., Witter, R. L., Jang, H. K., Ono, M., Mikami, T. and Silva, R. F. (1999): MDV glycoprotein D is expressed in the feather follicle epithelium of infected chickens. *Acta. Virol.*, 43: 159 – 163.

Perez, T. A. and Millan, A. D. A. (1994): Ia antigens are expressed on ATPase-positive dendritic cells in chicken epidermis. *J. Anat.* 184: 591 – 596.

Purchase, H. G. (1970): Virus-specific immunofluorescent and precipitin antigens and cell-free virus in the tissues of birds infected with Marek's disease. *Cancer Res.*, 30: 1898 – 1908.

Presland, R. B., Gregg, K., Molloy, P. L., Morris, C. P., Crocker, L. A. and Rogers, G. E. (1989): Avian keratin genes, I. A molecular analysis of the structure and expression of a group of feather keratin genes. *J. Mol. Biol.*, 209: 549 – 559.

Sevoian, M., Chamberlain, D. M. and Larose, R. N. (1963): Avian lymphomatosis. V. Air-borne transmission. *Avian Dis.*, 7: 102 – 105.

Sharma, J. M. and Burmester, B. R. (1982): Resistance to Marek's disease at hatching in chickens vaccinated as embryos with the turkry herpesvirus. *Avian Dis.*, 26: 134 – 149.

Spearman, R. I. C. and Hardy, J. A. (1985): Integument, Volume 3. London: Academic Press.

Tulman, E. R., Afonso, C. L., Lu, Z., Zsac, L., Rock, D. L. and Kutish, G. F. (2000): The genom of a very virulent Marek's disease virus. *J. Virol.* 74: 7980 – 7988.

Witter, R. L., Burgoyne, G. H. and Burmester, B. R. (1968): Survival of Marek's disease agent in litter and droppings. *Avian Dis.*, 12: 522 – 530.

Witter, R. L. (2001b): Protective efficacy of Marek's disease vaccines. *Curr. Top. Microbiol. Immunol.*, 255: 57 – 90.

Wozniakowski, G., Samorek-Salamonowicz, E. and Kozdrun, W. (2011): Rapid detection of Marek's disease virus in feather follicles by loop-mediated amplification. *Avian Dis.*, 55: 462 – 467.

Yu, M., Yue, Z., Wu, P., Wu, D. Y., Mayer, J. A., Medina, M., Widelitz, R. B., Jiang, T. X. and Chuong, C. M. (2004): The developmental biology of feather follicles. *Int. J. Dev. Biol.*, 48: 181 – 191.

Yue, Z., Jiang, T. X., Widelitz, R. B. and Chuong, C. M. (2005):Mapping stem cell activities in the feather follicles. *Nature*, 438: 1026: 1029.

Chapter 4

Infectious Bursal Disease

Infectious bursal disease (I.B.D.) is an acute and highly contagious infectious viral disease of young chickens (even day-old chicks) characterized mainly by severe lesions in Bursa of Fabricius resulting in to immuno-suppression. Infectious bursal disease virus (IBDV) produces clinical disease younger chickens while in older birds the infection remains essentially subclinical. Inoculation of IBD virus in other avian species fails to induce the disease (Mc Ferran, 1993). IBD virus is firstly reported from Gumboro, Delawara (U.S.A.) by Cosgrove in 1962 and hence named Gumboro disease. It is also called "Avian nephrosis". The disease is worldwide except New Zealand. In India, this disease entered via Nepal from China and later on spread over many states of the country.

The virus may show genetic drift and thus results in to increased virulence (Snyder *et al.*, 1992). Later on, very virulent (vvIBD) strains were emerged and rapidly spread over all Asia and other parts of the world. Now IBD is very endemic in India and many other parts of the world. It has been estimated that IBD has considerable socio-economic importance at the international level, as the disease is present in more than 95 per cent of the member countries of OIE/Office International des epizootics (Eterradossi, 1995). The European picture has also been enlightened for a decade because of the emergence of very virulent (vv) IBDV strains. These vvIBD virus strains have now spread all over the world.

Etiology

IBD virus is a small, non-enveloped virus, belonging to the family Birnaviridae, which is characterized by a bi-segmented dsRNA genome (Kibenge *et al.*, 1988). The virus has a single capsid shell of icosahedral symmetry composed of 32 capsomeres and a diameter of 60-70 nm.

The bi-segmented genome of virus has larger segment "A" (3400 base pairs/ bp) contains two open reading frames (ORFs). The larger ORF of segment "A" is monocistronic and encodes a poly-protein that is auto-processed after several steps in to mature VP2, VP3 and VP4 proteins (Kibenge *et al.*, 1997). Segment "A" also encodes VP5 protein, a short 17 Kilo Daltan (KD) protein, from a short, partially overlapping ORF (Mundt *et al.*, 1995). The smaller genome segment "B" (2800 bp) encodes VP1 protein which is the viral RNA Polymerase of 90 KD (Mullar and Nitschke, 1987; Spies *et al.*, 1987). The external surface of the virus is composed of trimeric sub-units formed by VP2 protein and the inner capsid is build of trimeric sub-units formed by VP3 protein (Lombordo *et al.*, 1999). The positively charged "C" terminus of VP3 protein might interact with the dsRNA genome (Hudson *et al.*, 1986; Bottcher *et al.*, 1997).

Broadly two serotypes are found *i.e.* serotype-1 and serotype-2 of which serotype -1 is more virulent. There are five structural proteins *viz.* – VP1, VP2, VP3, VP4 and VP5 in IBD virus. VP1 protein is RNA-dependent RNA polymerase of the virus (Kibenge and Dhama, 1997). VP2 is protective antigen and induces neutralizing antibodies in to host (Fahey *et al.*, 1989). VP3 is a group specific antigen which is recognized by non-neutralizing antibodies (Becht *et al.*, 1988, Oppling *et al.*, 1991). VP4 is a minor non-structural polypeptide (Azad *et al.*, 1987). VP5 protein has regulatory function which plays a key role in virus release and dissemination (Mundt *et al.*, 1997). Due to high mutation rate in the VP2 variable domain sequence, comparison of this region among strains offers the best evolutionary clue for IBD viruses as it is the proved molecular epidemiological tool for differentiation and identification of various strains of IBD viruses. A high rate of mutation occurs in RNA dependent RNA polymerase of IBD viruses that generates a genetic diversification which leads to emergence of the field viruses with new properties allowing them to persist even in immune populations. Till today, no biological or molecular structures have been identified as responsible factors of the virulence in IBD virus strains.

Hosts

Chicks (even day old) are susceptible but chicken between 2-7 weeks of age are most susceptible. However, from 9-days to 20 weeks old chickens are reported to be susceptible. After inoculation by allantoic sac route in a 10-days old chicken embryo, death of the embryo results between 3rd to 5th days showing edema and hemorrhages in subcutaneous tissues, around the joints, skull and liver. Liver of this embryo may show greenish appearance with necrotic spots.

Epidemiology

IBD virus survives for more than 4 months in litter, contaminated feed and water which are the chief sources of infection to the birds. The IBD viruses have also been isolated from mosquitoes which indicate the possible transmission of virus through mosquitoes (Canad. J. Comp. Med., 45: 315).

Pathogenesis

Pathogenesis can be defined as the method used by the virus to cause injury to the host with morbidity, mortality, disease or immuno-suppression as a consequence. The target organ for IBDV is the Bursa of Fabricius at its maximum development, which is specific source for B lymphocytes in avian species. Virus has a characteristic nature of immature B-lymphocytotropism. Bursectomy can prevent illness in chicks infected with virulent IBD virus (Hiraga *et al.*, 1994). The severity of the disease is directly related to the number of susceptible cells present in the Bursa of Fabricius, so the optimum age for maximum susceptibility is between 2-7 weeks, when the Bursa of Fabricius is at its maximum development. The susceptibility age window is broader in case of vvIBDV strains (Vandenberg *et al.*, 1991).

After oral infection or inhalation, the virus replicates primarily in B-lymphocytes and macrophages of gut associated lymphoid tissues (G.A.L.Ts.). Then virus travels to the Bursa of Fabricius via the blood stream, where further replication occurs. By 13-hours post-inoculation, most of the follicles in Bursa of Fabricius are found positive for IBD virus. By 16-hours post-inoculation, a second and more pronounced viraemia occurs followed by secondary replication in other target organs resulting in to disease and death (Muller *et al.*, 1979). Similar kinetics is observed for vvIBDVs but replication at each step is amplified. Actively dividing and surface immunoglobulin-M (Ig M) bearing B lymphocytes are lysed due to infection (Rodenberg *et al.*, 1994). Depletion of lymphoid cells in the Bursa of Fabricius after IBDV infection is due to both necrosis and apoptosis as macrophages produce tumor necrosis factor (T.N.F.). Although other different types of cells undergo apoptosis (Programmed Cell Death) but immature B and T lymphocytes are particularly susceptible to apoptotic cell death. This apoptosis of lymphoid cells leads to immuno-suppression. Antibodies can be detected after 3-days post-infection. The other important manifestation is "nephrosis" resulted due to deposition of immune complexes in renal glomeruli.

The disease is manifested mainly in two forms: 1). Clinical form, 2). Subclinical/ immuno-suppressive form.

1. Clinical Form

This form is usually seen in chickens of 3-7 weeks of age. In this form, remarkable enlargement of Bursa of Fabricius with gelatinous exudates deposited around the bursa within 3-5 days post-infection is seen. In case of very virulent strain infections, completely hemorrhagic bursa is found. After initial hypertrophy, the Bursa of Fabricius undergoes for atrophy within 7 days post-infection due to depletion of lymphoid cells in it. Kidneys swollen up and prominent ureters with urate deposition are other remarkable common observations. Linear or ecchymotic hemorrhages in thigh, breast, leg and wing muscles are pathognomonic lesions. Occasional hemorrhages in the mucosa of proventriculus-gizzard junction and on heart surface are found. Initially, spleenomegaly with gray foci scatered on it followed by atrophy of spleen may be seen. Sometimes, hepatomegaly is also developed.

2. Subclinical/Immuno-suppressive Form

This form of disease is seen particularly in broilers complicated with secondary bacterial infections of various types. Variant IBDV serotype -1 causes severe immuno-suppression but no mortality. However, standard serotype-1 and very virulent serotype-1 IBDV cause immuno-suppression as well as mortality. In this form, sporadically mild bursal lesions and atrophy of bursa are seen. Atrophy of Bursa of Fabricius is most prominent at 23rd day post-infection.

Signs and Symptoms

In India, due to IBD infection morbidity varies between 80-100 per cent while mortality varies between 40-80 per cent. 3-6 weeks old chicks suffer more with 100 per cent mortality. Watery and white diarrhea, calmly sited birds with closed eyes, trembling and shivering with initially raised body temperature is observed. Later on, the temperature may be normal or subnormal. Pecking of vent is alarming sign. Swollen Bursa of Fabricius is common feature. Due to immuno-suppression, any pathogen can infect the affected bird and the signs and symptoms of respective developed diseases can be observed.

Diagnosis

Diagnosis of IBD is based on clinical signs, age associated high mortality pattern and gross lesions in affected birds. 50 per cent of the dead birds show coagulative degeneration of hepatocytes. Kidneys reveal the presence of eosinophilic casts in tubular lumen and necrosis in the epithelium of proximal convoluted tubules. An increased number of macrophages are found in various affected organs (Tanimura *et al.*, 1995). Acute disease is characterized by disseminated hemorrhages in various muscles and organs probably related to an impairment of the clotting mechanisms as thrombocytes are also target cells for IBD virus (Skeeles *et al.*, 1980). Severe depletion of lymphoid cells is observed not only in the Bursa of Fabricius but also in the non-bursal lymphoid tissues.

Virus isolation and identification is confirmatory diagnosis. Virus isolation is done in embryonated chicken eggs. Bursa of Fabricius in acute phase of disease is the most suitable source of virus. A 10-11 days old chicken embryo is inoculated with morbid material through CAM (Chorio-allantoic membrane) route, embryo dies within 3-5 days after inoculation showing mottled greenish liver, mottled kidneys and congested lungs. Edematous embryo, hemorrhages in skin, joints and toes are seen. Immuno-fluorescence, agar gel precipitation test (A.G.P.T.), virus neutralization test (V.N.T.), enzyme linked immunosorbent assay (E.L.I.S.A.), reverse transcription-polymerase chain reaction (R.T.-P.C.R.) and restriction fragment length polymorphism (R.F.L.P.) can be used as newer and modern tool for IBD diagnosis.

Treatment

No specific treatment is available for IBD. Only by vaccination this disease can be prevented. However, secondary bacterial complications should be treated with suitable antibiotics and other supportive therapies.

Vaccines and Vaccination

The vaccines and vaccinations against IBDV largely depend upon the principles of humoral immunity. This humoral immunity is based on active and passive protection through own body generated antibodies in adult birds (Broilers/Layers/Breeders) while in young chicks it is achieved by maternally derived antibodies (MAbs) respectively. These MAbs are passively transferred from well immunized hens to the progeny chicks via egg yolk and it remains present in sufficient concentration through 3rd weeks of age of birds. The level of antibodies and MAbs in adult birds and chicks respectively are measured by ELISA in terms of titers. An ELISA titer of 500 or more is thought to be protective for the individual bird. It has been observed that the antibody titers of day old chicks are positively related to the dam's titer. MAbs are protective against IBDV and also against immune-suppressive effects of the virus. Therefore, on this basis, the MAbs titers of the progeny chicks are determined. This provides the base for predicting the susceptibility age and hence the suitable timing for vaccination of chicks against IBDV. Chicks with high level of MAbs titer remains protected from immune-suppression and bursal atrophy in early age of life. The level of transfer of MAbs from dam to a day old progeny chick is about 40-45 per cent of the antibody titer of dam and the minimum protective ELISA titer is 500 as a standard. MAbs abolishes in a way as radioactive substances decay *i.e.* according to half life ($T_{1/2}$) theory. The $T_{1/2}$ of MAbs is 3-3.5 days in broiler, 4 days in breeder and 4.5-5 days in layer progeny chicks. $T_{1/2}$ (Half life) is the time period during which the concentration of MAbs becomes half of the initial concentration in circulation. The presence of high titer of MAbs (>500 ELISA titers for intermediate vaccines and >200 ELISA titers for mild vaccines) in progeny chicks interfere the early vaccination as it can neutralize the vaccine antigen resulting into vaccination failure, while low titers of MAbs induce susceptibility of the chicks to contract IBD infection resulting in to immuno-suppression.

Types of Vaccines

Broadly two types of vaccines are there. 1)- Live attenuated vaccines and 2)-Killed/Inactivated vaccines. Live attenuated IBD vaccines contain live immunogenic viruses and are of three categories: A)-Mild type vaccines, 2)-Intermediate type vaccines and 3)- Hot/Strong type vaccines. These three categories of live IBD vaccines are based on the number of their passages, degree of pathogenic residuality and ability to break through the MAbs titers.

Mild type of vaccine is generally used for priming vaccination. It is used in chicks when MAbs ELISA titers come to 200 or below. Then only this type of vaccine reveals sero-conversion and show active immunization, otherwise the vaccine will be neutralized without showing immunogenesis and dangerously lowers the level of circulating MAbs, thus increasing the chances of getting field infection as the window of infection susceptibility is increased in this situation.

Intermediate type of live attenuated IBD vaccine is very usefull. It breaks through the MAbs ELISA titer of 500 (which is also protective to young chicks). It is able to provoc immune response and narrows down the infection susceptibility window reducing the entry of field virus in the chicks of flock in question. Unfortunately, most intermediate IBD live attenuated vaccines were found as inadiquate for interfering with vvIBDV that could break through higher MAbs levels.

Hot/Strong type live attenuated IBD vaccine has more degree of residual pathogenecity. So, it breaks through the MAbs ELISA titer of >500. It is used when high ELISA MAbs titer is present and even birds have more mortality. It checks the mortality but induce immuno-suppression. However, the use of less attenuated (Hot/Strong) vaccines, even with an acceptable reduction of mortality, is dangerous as these vaccines induce immunosuppression and carry the risk of reversion to virulence.

Killed/inactivated vaccines are oil emulsion vaccines, which generally used in breeder flocks in order to raise high antibodies titer in hens resulting in to higher proportionate transfer of MAbs in progeny chicks. These inactivated vaccines are usually injectables.

Characteristics of Live Attenuated IBD Vaccines

Vaccine Strains Characteristics

	Vaccines	Mild Type	Intermediate Type	Hot Type
1	Microscopic bursal lesions	+	++	+++
2	Serological response	+	++	+++
3	Interference by MAbs	+++	++	+
4	Bursa/body weight ratio	+++	++	+
5	Depression of ND titers	+	++	+++
6	Immunizing dose	+++	++	+
7	Breaking through of MAbs ELISA titers	+ (200 or less)	++ (500)	+++ (>500)

Mild live attenuated vaccines are more readily neutralized by MAbs resulting in higher bursal/body weight ratios and require a high dose to immunize the birds. It can only break through an ELISA titer of 200 or less to provoc immunogenesis. This type of vaccine is generally used for priming before an inactivated/killed vaccine administration.

Intermediate type of live attenuated IBD vaccines show intermediate characters and can be able to induce the immunogenesis at 500 ELISA titers without being much neutralized by MAbs. Hot/Strong type live attenuated IBD vaccines usually cause more bursal lesions, higher serological response, greater depression in New Castle disease (ND) antibody titers and inducing immunogenesis in the presence of even higher titers of MAbs *i.e.* >500 ELISA titers without being much neutralized. In those areas of the world, where highly pathogenic/very virulent IBDV occurs, hot/strong vaccines are required to prevent severe mortality. Selection of the age at which this hot/strong vaccine has to be administered in drinking water is based on the rate of decline in MAbs titer in progeny chicks.

Recently a new concept, which consists of the in ovo inoculation of a virus-antibody complex vaccine has emerged (Haddad *et al.*, 1997). This novel technology utilizes specific hyper-immune neutralizing antiserum (or virus neutralizing factor/VNF) with a vaccine virus under conditions that are not sufficient to neutralize the vaccine virus but which are sufficient for delaying the pathological effects of the vaccine alone. This allows young chicks to be vaccinated more effectively in the presence of passive immunity even with a strain that would be too virulent for use in ovo or at hatching. Another novel vaccine is recombinant vaccine with fowl pox virus (Bayliss *et al.*, 1991), Herpes virus of turkey/HVT (Tsukamoto *et al.*, 1999) or fowl adenovirus (Sheppard *et al.*, 1998) expressing the VP2 protein of IBDV, might prove to be a powerful tool in the near future in priming an active immune response.

So, the choice of vaccine depends upon the virulent challenge of field virus, Chick's MAbs titer and age at infection. The uniform level of MAbs, non-uniform level of MAbs, field challenge, type of vaccine and breed of the birds (chicken) are determinant factors for successful IBD vaccination. Moreover, better identification of the protective criteria and differentiation between active and passive immunity might be of considerable help in the establishment of vaccination schedule.

a) Effect of Uniform MAbs Titer Level

The MAbs serve to check the early exposure to IBD virus and also block active immunogenesis with less strong live vaccines.

But it is clear that if the levels of MAbs are almost uniform in terms of titer, successful complete flock immunization can be achieved with a single vaccination and flock immunity can be established.

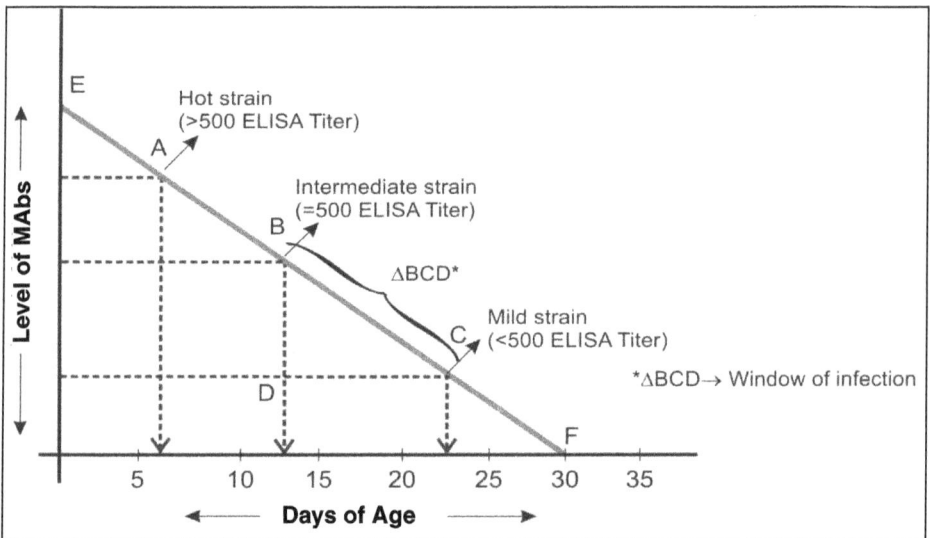

Figure 1: Uniform MABs Level Situation (E to F).

b) Effect of Non-uniform MAbs Titer Level

In many situations, the MAbs titers are not uniform in the chicks of a flock. The reasons of which may be:

1. Mixing of chicks in a hatchery from different breeder flocks, which may be of different breeds, age and/or using different vaccination programmes in breeder flocks.

2. Individual hereditary variational changes to transfer variable amount of MAbs by the breeder hen to each chick.

3. Differences between the hens antibody titers resulting in to difference in amount of MAbs transferred to progeny chicks.

A single vaccination given to such flock as with variable titer of MAbs results in partial flock immunization. So, multiple applications of IBD vaccines are often needed. The goal of multiple vaccination programme is to successfully immunize each chick in the flock *i.e.* stimulating an active immune response in each chick of the flock in question.

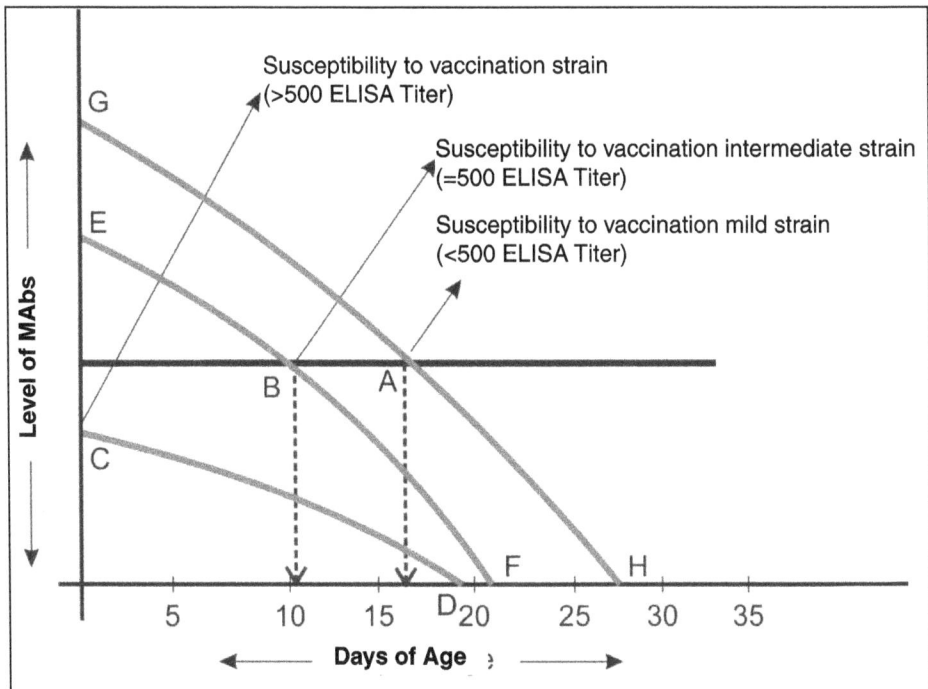

Figure 2: Effect of Non-Uniform MABs Level (CD, EF and GH).

c) Effect of Field Challenge

The pathogenecity of field virus in the presence of MAbs affects vaccination timing and choice of vaccine. If the flock has to be vaccinated earlier in the presence of high field virus challenge, the MAbs ELISA titer of 500 or more should be broken through by intermediate or hot/strong live vaccines, respectively.

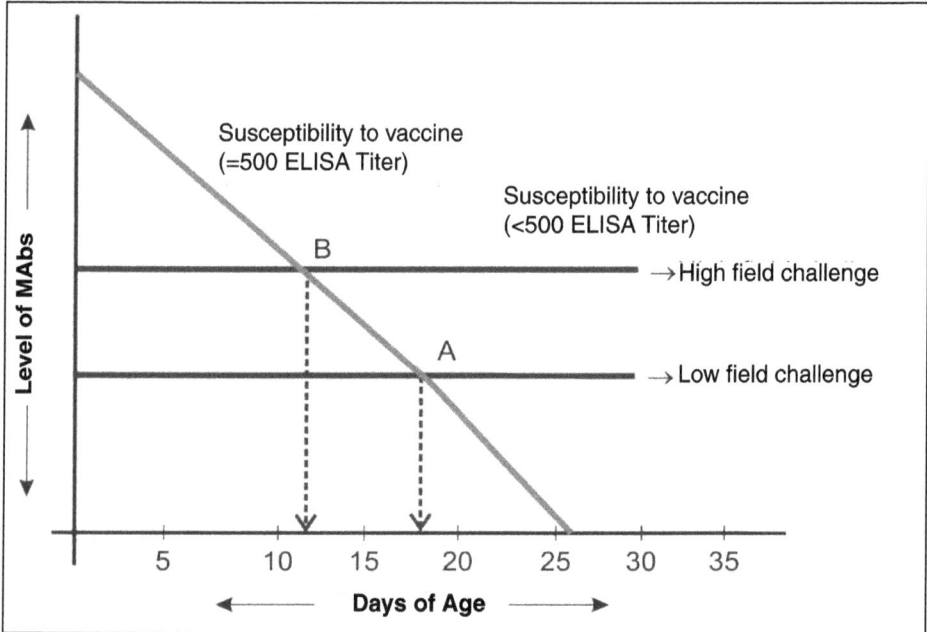

Figure 3: Effect of Field Challenge.

But if there is low field challenge, the vaccination with mild strain live vaccine should be practiced, as mild live vaccine is able to break through 200 or less ELISA titer of MAbs.

d) Effect of Type of Vaccines

The type of IBD live vaccines (hot/strong, intermediate or mild) used can affect proper vaccination timing and schedule due to differences in the ability of vaccines to successfully immunize all the chicks of a flock in the presence of MAbs.

Vaccinations with mild types of vaccines are needed to be practiced at an older age than that with intermediate or hot/strong vaccines in the presence of high MAbs titers because mild vaccines will be readily neutralized by higher titers of MAbs of the chicks.

e) Effect of Breed

The half life $(T_{1/2})$ of MAbs *i.e.* the time period during which the initial concentration of MAbs in circulation becomes half, is different in broiler, breeder and layer chicks which is 3-3.5 days, 4 days and 4.5-5 days respectively. This largely affects the vaccination timing and schedule of the chicks of various breeds. It makes difference that broiler breeds of chicks should be vaccinated earlier than that of breeder and layer breeds of chicks.

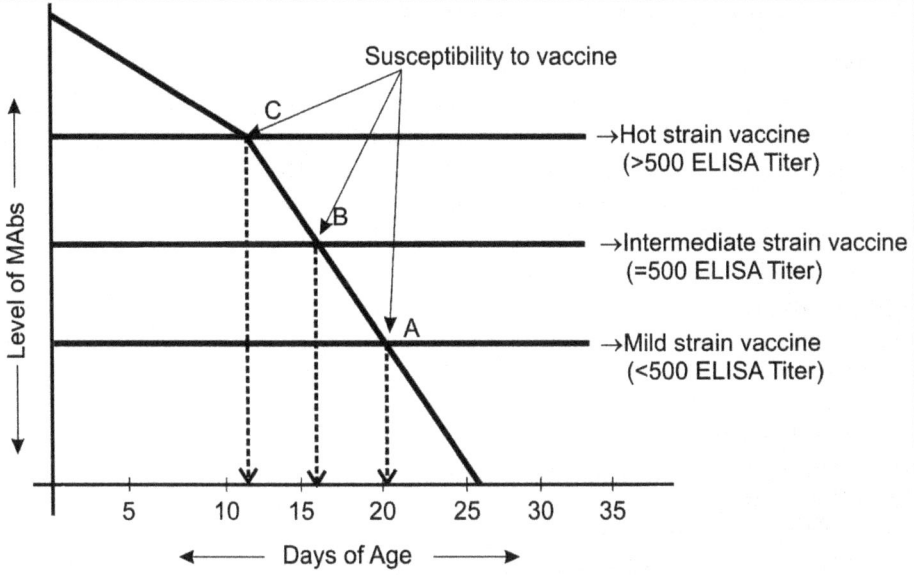

Figure 4: Effect of Type of Vaccines.

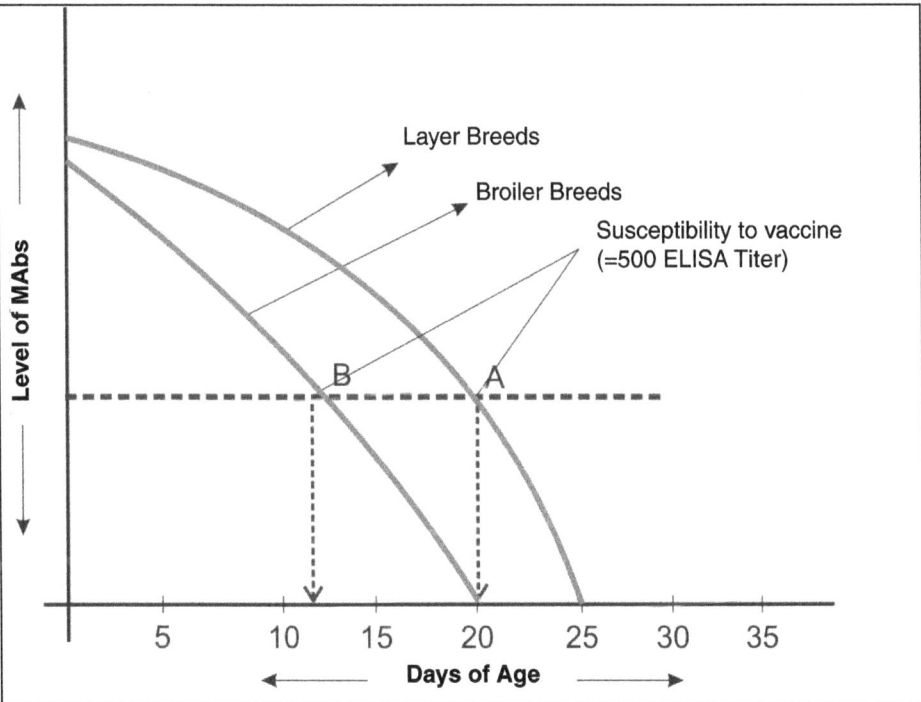

Figure 5: Effect of Breeds.

Layer/Breeder Vaccination Programmes

Programme – 1

The objective of this programme is to protect pullets from clinical and immunosuppressive infectious bursal disease (IBD). This only live vaccination programme is for use in layer and breeder flocks with variable MAbs titers. In breeders, this programme is only practiced when low MAbs titers in progeny chicks are needed. Successful live IBD vaccination protects pullets from IBD.

In the area of low field virus challenge, one successful live mild vaccination at 30-35 days, when all circulating MAbs in pullets are depleted, may be sufficient and adequate. Mild attenuated live vaccines can be used in low challenge areas to avoid the potentially negative/damaging effects of intermediate and hot/strong live attenuated vaccines on the Bursa of Fabricius. Variable MAbs profiles in young pullet flocks result in partial flock immunization needing at least two live mild vaccinations necessary to provide complete flock protection/immunization. Early vaccination before 7 days of age is justified where variable levels of MAbs titer and early field virus challenge exist. Mild live vaccines are neutralized by even relatively low level of circulating MAbs making later vaccination timings after 14 days necessary. For breeders, live vaccinations alone don't promote significant MAbs titer levels in the progeny chicks. Hot/strong vaccines are rarely used in layers and breeders except in the locations where very virulent infectious bursal disease (vvIBD) is present. Generally, intra-ocular vaccination route is practiced (before 20 days of age of birds) followed by drinking water vaccination because the advantages of using intra-ocular applications are that fewer or no bird(s) are missed for the receiving of vaccine dose and a more uniform dosage of vaccine is administered resulting in to provoking uniform antibody titer. Intra-ocular vaccination is practiced where labour cost is low. Coarse spray vaccination is not much efficacious in comparison to intra-ocular vaccination followed by drinking water vaccination. Layer breeds are more susceptible to the damaging effects of live IBD vaccines. So, hot/strong vaccines are only used when there are chances of high challenge of vvIBD virus resulting in to chances of expected high mortality.

Programme – 2

The objective of this vaccination programme is to protect pullets from clinical and immuno-suppressive IBD and to stimulate high MAbs titers in the progeny chicks by live and inactivated combined vaccination programmes.

Firstly, breeders are primed with live attenuated vaccine followed by inactivated vaccine. This priming is done between 8 – 14 weeks of age of pullets by intermediate live vaccine because mild vaccines are less affective in the pullets aging 10 -12 weeks. The exact timing of this intermediate type vaccine administration is calculated on the basis of ELISA titer of circulating MAbs in pullets. For determination of mean ELISA titer and hence vaccination timing of the flock, at least 25 random sera samples amongst the day old chicks should be taken. These sera samples are obtained from blood samples directly taken from the heart of the day old chicks through cardio-puncture.

Inactivated vaccine is generally administered 4 weeks prior to the onset of egg production to allow sufficient time period for full development of immune response. Peak antibody titer is attained by dam within 4–6 weeks post vaccination. Some inactivated vaccines contain multiple strains of the virus which also include variant field virus strains to broaden the spectrum of protection by MAbs in progeny chicks. The residual tissue reactions may be seen even after several weeks post vaccination. Breeder's dam IBD antibody titer may drop after 40-45 weeks of age of birds. The antibodies produced in active immunization don't follow the rule of radio-active decay in terms of half life $(T_{1/2})$ as in case of passively transferred circulating MAbs in chicks and pullets. So, at 35–40 weeks of age of breeder/layer dams, ELISA test is recommended to determine the antibody titer and at proper time (when ELISA titer is dropped down significantly) another booster dose of inactivated vaccine is recommended. This booster dose is sufficient for rest of the whole active optimum production life. The handling of lay hen/dam should be very gentle. Vaccination during this period may result in transient decrease of egg production which automatically recovers after a short time period. Inactivated vaccines should be administered as injections and not by intra-ocular or in drinking water routes.

Programme – 3

The objective of this vaccination programme is to protect pullets from clinical and immuno-suppressive IBD and to stimulate high MAbs titers in progeny chicks through use of only inactivated vaccines.

The programme is an alternative to mid lay inactivated IBD vaccination thus avoiding the handling and vaccine stress during mid lay period. An inactivated vaccine is advised to be administered at the age of 10–12 weeks replacing the intermediate live vaccination as practiced in programme-2. This programme offers high and uniform antibody titer in dams without mid lay period handling stress. Thus, the use of two successive inactivated vaccines incorporating variant strains also, makes possible not only full protection to breeders/layers but also high MAbs titer levels in progeny chicks. These two inactivated vaccines should be given 4 weeks apart and last administration should be 4 weeks prior to the onset of egg production *i.e.* 1st inactivated vaccination at 8th week of age and 2nd inactivated vaccination at 12th week of age of birds. This will be sufficient for whole active optimum egg production life of the dams.

Broiler Vaccination

Programme – 1

The objective of this programme is to protect broiler flocks of less than 3 weeks of age against early field virus infection, which can result in permanent immuno-suppression. Here the broiler progeny chicks may have low or variable MAbs titers.

To prevent the chicks having low or no circulating MAbs (even if they represent a small percentage of the flock) from replicating field virus, hatchery vaccination can be adopted justifiably. Hatchery live IBD vaccine may be administered with Marek's disease vaccine. In order to accomplish complete flock immunization, multiple applications of live IBD vaccines are necessary due to variable titers of MAbs in chicks of the flock. The goal of this multiple vaccination is to successfully immunize each chick in the flock. Successful immunization is defined as stimulating an active immune response in all the birds of the flock. Intermediate live IBD vaccines are generally used for field vaccinations because they can immunize the chicks in the presence of significant MAbs titers and lacking the quality of high pathogenicity of hot/strong vaccines.

Live variant strains vaccines have been used in the areas where variant field virus strains are established and standard strain vaccines have not been proved as protective. Broilers generally experience earlier field challenge than commercial layers and breeders because they are raised in environments more heavily contaminated with field viruses which is due to the result of shorter intervals between flocks, higher stocking densities, reuse of litter, less stringent disinfection and lapsed biosecurity measures. Here again the use of intra-ocular applications of live vaccine is more appropriate as in this way very few or no bird is missed in getting the vaccine. However, most hatchery IBD vaccines are not promising to protect the broilers because of high titers of circulating MAbs in progeny chicks and therefore, the vaccine may be neutralized. So, only use of specially designed tailor made vaccines should be administered in hatchery. The IBD vaccine designed for use in combination with Marek's disease vaccine is only be used in hatchery. This specially designed tailor made IBD vaccine doesn't interfere with Marek's diaease immunity. Live IBD variant strains vaccines should not be introduced in to the areas where these variant strains of IBD virus are not in existence or established.

Programme – 2

The objective of this vaccination programme is to protect chicks of less than 3 weeks of age against early field virus infections resulting in to permanent immuno-suppression and later infections at more than 3 weeks of age which can result in clinical disease and temporary immuno-suppression.

The alternative of hatchery vaccination is field vaccination. In order to accomplish complete flock immunization, multiple applications of live IBD vaccines specially intermediate type vaccines are often needed because of variable titers of MAbs in the flock. The goal of multiple vaccination programme is to successfully immunize *i.e.* creating active immunity in each chick of the flock.

Multiple vaccination with live intermediate IBD vaccines have been used to control mortality associated with the very virulent form of clinical IBD when used in conjunction with high level of sanitation and biosecurity measures.

Programme – 3

The objective of this vaccination programme is to protect broilers against field infections due to very virulent IBD viruses (vvIBDVs) which result in clinical disease, high mortality and immuno-suppression.

The hot/strong strains IBD vaccines have been used in areas experiencing outbreaks of very virulent forms of clinical IBD. These vaccine strains can immunize chicks in the presence of high titers of MAbs and control the high mortality associated with this very virulent form of IBD (vvIBD).

The intermediate live attenuated vaccine is usually given once followed by hot/strong live attenuated vaccine. The intermediate vaccine is used to immunize those birds with low (ELISA titer of 500) titers of MAbs which would be susceptible to the bursal damage caused by some hot/strong vaccines. The decision is to use intermediate type or mild type vaccine firstly in this programme is to be decided on the basis of ELISA titer of MAbs circulating in chicks of the flock in question. If the general ELISA titer distribution is around 500, use of intermediate live attenuated vaccine is recommended but if this ELISA titer is 200 or less, mild live attenuated vaccine is preferred. Hot/strong vaccine may shift the resident virus population in a house from the pathogenic field virus to one of the vaccine origin. Intermediate live attenuated vaccines may be used in subsequent flocks to replace the hot/strong vaccine strain.

Changes in the broiler performance often occur over 3–5 grown out batches as the pathogenic virus is displaced from the house by the vaccine virus strains. The live attenuated intermediate vaccines should be administered intra-ocularly to immunize each chick/bird of the flock. The use of hot/strong live vaccines can result in immuno-suppression and hence production losses.

REFERENCES

Azad, A. A., Jagdish, M. N., Brown, M. A. and Hudson, P. J. (1987). Deletion mapping and expression in E. coli of the large genomic segment of a birnavirus. *Virology*, 161, 145 – 152.

Becht, H., Muller, H. and Muller, H. K. (1988). Comparative studies on structural and antigenic properties of two serotypes of infectious bursal disease virus. *Journal of General Virology*, 69, 631 – 640.

Bottcher, B., Kiselev, N. A., Stel'Mashchuk, V. Y., Perevozchikova, N. A., Borisov, A. V. and Crowther, R. A. (1997). Three – dimentional structure of infectious bursal disease virus determined by electron cryomicroscopy. *Journal of Virology*, 71, 325 – 330.

Fahey, K. J., Erny, K. and Crooks, J. (1989). A conformational immunogen on VP2 of infectious bursal disease virus that induces virus-neutralizing antibodies that passively protect chickens. *Journal of General Virology*, 70, 1473 – 1481.

Hiraga, M., Nunoya, T., Otaki, Y., Tajima, M., Saito, T. and Nakamura, T. (1994). Pathogenesis of highly virulent infectious disease virus infection in intact and burectomized chickens. *Journal of Veterinary Medical Sciences*, 56, 1057 – 1063.

Hudson, P.J., McKern, N.M., Power, B.E. and Azad, A.A. (1986). Genomic structure of the large RNA segment of infectious disease virus. *Nucleic Acid Research*, 14, 5001 – 5012.

Kibenge, F. S. B. and Dhama, V. (1997). Evidences that virion-associated VP1 of avibirnaviruses contains viral RNA sequences. *Archives of Virology*, 142, 1227 – 1236.

Kibenge, F.S.B., Dhillon, A.S. and Russel, R.G. (1988). Biochemistry and Immunology of infectious disease virus. *Journal of General Virology*, 69, 1757 – 1775.

Kibenge, F.S.B., Qiana, B., Cleghorn, J.R. and Martin, C.K. (1997). Infectious bursal disease virus polyprotein processing doesnot involve cellular proteases. *Archives of Virology*, 142, 2401 – 2419.

Lombardo, E., Maraver, A., Casten, J.R., Rivera, J., Fernandez-Arias, A., Serrano, A., Carrascosa, J.L. and Rodriguez, J.F. (1999). VP1 the putative RNA dependent RNA polymerase of infectious bursal disease virus, forms complexes with the capsid protein VP3, leading to efficient encapsidation into virus-like particles. *Journal of Virology*, 73, 6973 – 6983.

McFerran, J.B. (1993). Infectious bursal disease. In J.B. McFerran and M.S. McNulty (Eds.), *Virus infections of birds* (pp. 213 – 228).

Muller, H. and Nitschke, R. (1987). The two segments of the infectious bursal disease virus genom was circularized by a 90,000-Da protein. *Virology*, 159, 174 – 177.

Muller, R., Kaufer, I., Reinacher, M. and Weiss, E. (1979). Immuno-florescent studies of early virus propagation after oral infection with infectious bursal disease virus (IBDV). *Zentralblad Veterinarmedicine* [B], 26, 345 – 352.

Mundt, E., Beyer, J. and Muller, H. (1995). Identification of a novel viral protein in infectious bursal disease virus-infected cells. *Journal of General Virology*, 76, 437 – 443.

Mundt, E., Kollner, B. and Kretzschmar, D. (1997). VP5 of infectious bursal disease virus is not essential for viral replication in cell culture. *Journal of Virology*, 71, 5647 – 5651.

Oppling, V., Muller, H. and Becht, H. (1991). Heterogenecity of the antigenic site responsible for the induction of neutralising antibodies in infectious bursal disease virus. *Archives of Virology*, 119, 211 – 223.

Rodenberg, J., Sharma, J., Belzer, S. W., Nordgren, R. M. and Nagi, S. (1994). Flow cytometric analysis of B cell and T cell subpopulations of virus. *Avian Diseases*, 38, 16 – 21.

Skeeles, J. K., Slavik, M., Beasley, J. N., Brown, A. H., Meinecke, C. F., Maruca, S. and Welch, S. (1980). An age-related coagulation disorder associated with experimental infection with infectious bursal disease virus. *American Journal of Veterinary Research*, 41, 1458 – 1461.

Snyder, D.B., Vakharia, V.N. and Savage, P.K. (1992). Naturally occuring-neutralizing monoclonal antibody escape variants define the epidemiology of infectious bursal disease viruses in the United States. *Archives of Virology*, 127, 89 – 101.

Spies, U., Muller, H. and Becht, H. (1987). Properties of RNA polymerase activity associated with infectious bursal disease virus and characterization of its reaction products. *Journal of General Virology*, 69, *Virus Research*, 8, 127 – 140.

Tanimura, N., Tsukamoto, K., Nakamura, K., Narita, M. and Maeda, M. (1995). Association between pathogenicity of infectious disease virus and viral antigen distribution detected by immuno-chemistry. *Avian Diseases*, 39, 9 – 20.

Vandenberg, T. P., Gonze, M. and Meulemans, G. (1991). Acute infectious bursal disease in poultry: isolation and characterisation of a highly virulent strain. *Avian Pathology*, 20, 133 – 143.

Chapter 5

Infectious Bronchitis

Infectious bronchitis (IB) is an infectious viral disease of poultry caused by gamma-Coronavirus and characterized by respiratory symptoms, drop in egg production, production of soft shelled or misshapen eggs and decrease in body weight gain. Infectious bronchitis virus (IBV) is firstly isolated in USA during 1930. The virus belongs to genus gamma-Coronavirus, subfamily Coronavirinae and family Coronaviridae of order Nidovirales. Some mammalian Coronaviruses are alpha and beta coronaviruses. Novel genus is delta-Coronavirus causing infection in wild birds and pigs (Woo *et al.*, 2012). Recently, the Coronavirus study group of the International Committee for Taxonomy of Viruses (ICTV) proposed three genera: Alpha, Beta and Gamma Coronaviruses, further divided into subgroups 2a, 2b, 2c and 2d and 3a, 3b and 3c replacing the traditional groups 1, 2 and 3 (Woo *et al.*, 2009). Coronaviruses are enveloped and have a non-segmented, positive sense, single strainded RNA genome.

The Coronaviruses present a large surface of protein "S" (Spike protein), responsible for the adsorption of the viruses on host cells, inducinging the fusion of viral envelop with host cells plasma membranes for subesequent release of viral RNA in to cytoplasm. There are three other stuctural proteins N, E and M. S = Spike glycoprotein which results virus attachment and neutralizaton epitopes, N = Nucleoprotein which surrounds and protect viral RNA genome, E = Envelop (small membrane) protein which is important for virus assembly and M = Membrane protein which is integral membrane protein.

The main componental protein is "S" protein/Spike protein which due its position in the viral envelope, determines the "crown shape" of these viruses as seen under electron microscope. The "crown" is characteristic of viruses of Coronaviridae family. This "S" proein occurs as a dimer or trimer (Lewicki and Gallagher, 2002) and can be cleaved in to two subunits S1 and S2. S1 is amino terminal component while S2 is carboxy terminal component. This "spike protein" is important in determining virus specificity and it is also involved in the pathogenicity of the virus (Zeng *et al.*, 2006).

IBV serotype classification is based on S1 spike protein differences. Many serotype present 20 – 25 per cent differences in S1 amino acids while others >50 per cent. Differences in other viral proteins rarely increase 15 per cent. Thus, generally immunity against one serotype provides poor protection against other serotypes, considering that cross-protection decreases when differences in "spike proteins" (S1 and S2) increase higher than 5 per cent.

Epidemiology

IBV is distributed worldwide. Chickens are the most important natural hosts. Chickens of all ages may be affected. In addition to chickens, avian Coronavirus has been found in many species of birds including turkeys, pheasents, pigeons, ducks, teals, peafowls, red knots, goose, whoopers, swan, quail, thrush, minia and bulbuls (Britton and Cavanagh, 2007, Cavanagh *et al.*, 2002). The most sever clinical signs are seen in chickens below 6 weeks of age. The morbidity rate is high but mortality rate depends upon the age of birds when infected and the presence of secondary invading organisms like E. coli. Mrtality by alone IBV results as approximately 10 per cent but when complicated with other microbes, it reaches up to 50 – 100 per cent. IB is transmitted through air borne route, direct chicken to chicken contact and indirectly through equipements, egg packing materials, mannure, farm visitors and drinking water. IB can occur in any season but it occurs more in winter. Carrier birds may carry the infection up to 2 months post infection.

Pathogenesis

The IBV replicates in respiratory epithelial tissues as well as in other epithelial tissues of the organs like kidneys, gonads, oviducts and Bursa of Fabricius (Covanagh, 2007a). IBV initially infects and replicates in the upper respiratory tract resulting in loss of protective cell linings in the sinuses and trachea. After a brief viremia, the virus can be detected in kidneys, reproductive tracts, gonads and ceacal tonsils.

The principal clinical manifestations are respiratory signs that develop after infection of respiratory tract tissues following inhalation or ingestion. Sneezing is important fact in the transmission. Infection of oviduct can lead to permanent damage in immature birds. In laying hens infection leads to ceasation of egg laying and/or production of thin walled or misshapen egg shells with loss of shell pigmentation. Some strains of IBV reffered as nephropathogenic strains are known to cause lesions in the kidneys. Renal damage associated with different IBV strains are more important in broilers. The disease results in acute nephritis and urolithiasis resulting heavy mortality (Cavanagh and Gelb, 2008). After apparent recovery, chronic nephritis can result in death of the birds. IBV has also been reported to produce disease of proventriculus (Yu *et al.*, 2001). Vaccine and field strains of IBV may persist in the ceacal tonsils of the intestinal tract and be excreted in feces for weeks or longer in even clinically normal chickens (Alexander *et al.*, 1978). IBV doesnot infect human beings, so it is not zoonotic. IBV incubation period is relatively shorter (18–36 hours). So, the disease can be spread in entire flock with in 1–2 days after contracting infection.

The virus initially infects the URT (upper respiratory tract) and after 3 days of inoculation, the highest titers of virus can be found in trachea (Cavanagh, 2003), which may persist for up to 5 days depending upon the virus strain. IBV encoded proteins modulate and induce apoptosis, sometimes directly contributing to virus pathogenecity or inhibits apoptosis as the prevention of the early death of infected cells allows viral replication resulting in high titers of the virus in the infected hosts (Enjuanes *et al.,* 2006). Such balance of pro or anti-apoptotic molecules affects cell survival in the early stages of infection. Interestingly, pro-apoptotic molecules target specific tissues influencing the clinical manifestations of the disease, where as the anti-apoptotic molecules maitain the virus population for longer period allowing mutation and adaptation in the virus. This leads to the emergence of new variant virus strains. Thus, IBV has an ability to change its genomic make up and immunogenicity rapidly.

Signs and Symptoms

IBV affects the respiratory and uro-genital tracts of the birds. Birds of all age groups are susceptible to the infection but the clinical signs may vary. The firstly recognized and most conspicuous signs are respiratory signs hence, the name infectious bronchitis has been given. However, the pathogenecity of the virus for the oviduct in very youg chicks or birds in production is often more important. The kidneys may also be infected. Following signs may be seen:

1. Young chickens are depressed and huddle under the heat source.
2. Gasping, coughing, tracheal rales and nasal discharges are seen.
3. Birds in lay have marked drop in egg production and production of poor shell quality eggs are resulted.
4. The external and internal quality of eggs may be affected resulting in misshapen or soft shelled eggs with watery containt.
5. The hatchability rate of eggs may be affected adversely.
6. When the kidneys are affected, increased water intake, depression, scouring and wet litter are commonly observed.

Diagnosis

Diagnosis is based on clinical signs and laboratory findings. The novel method for detection of viral genomes is reverse transcriptase – polymerase chain reaction (RT-PCR). For nucleic acid sequencing R.F.L.P. and and for IBV typing real time-PCR is adopted.

For the detection of antibodies HI, ELISA, AGPT and VNT are adopted.

One of the most reliable methods of diagnosis is virus isolaton and identification. IBV is most sucessfully isolated from infected tissues of tracheal mucosa and lungs (with in 1 week following infection), kidneys, oviducts, proventriculus and ceacal tonsils. Virus isolation is usually done in 9–10 days aged embryonated specific pathogen free (SPF) eggs. Typical lesions occuring at 5-7 days post-inoculation are curling and dwarfism of the embryos, clubbing of down, red or hemorrhagic embryos and possible white urate deposition in the uraters/kidneys. Tracheal rings may also be used for isolation of IBV. Tracheal damage occurs with in 48–72 hours post-inoculation.

Protectotype

IBV strains with different antigenic or genetic features may still croos-protect in vivo. Cross-protective strains of IBV are called as "protectotypes". The most commonlly known IBV serotype is the "Massachusetts serotype" which represents the most important "protectotype" because it has an ability to cross-protect against a number of viruses belonging to different serotypes or genotypes of IBV.

Economic Consequences of IB

Broilers may perform badly due to poor feed conversion resulting in to reduced body weight gain. Secondary bacterial infections may lead to increased number of condemnation at slaughter. In layers/breeders IBV badly affects egg production. Layers and breeders may be infected in early ages with a very virulent IB virus (vvIBV) resulting in to permanent damage in oviducts. These birds may mature like other normal hens but produce no egg at maturity. These are called as "false layers" which consume feed, water and facilities but don't produce eggs. During laying period infection, egg production is adversely decreased in terms of both quality and quantity. The hatchability of these eggs is markedly reduced.

Treatment

There is no specific treatment. Secondary bacterial infections should be treated accordingly.

Prevention and Control

Biosecurity is an effective measure for prevention and control along with suitable vaccination programme. Basic management practices such as limited controlled site access, separate footwear and equipements for each site/house and footbaths at the entrance to sites/houses minimize the risk of introducing IBV infection. Dry cleaning of houses by removing and disposing of all the organic materials from houses is recommended followed by wet cleaning of houses under high pressure of water and detergents is advisable. Disinfection using formaldehyde, chlorine releasing chemicals or quartenary ammonium compounds are essential of effective biosecurity.

Immunological Considerations

Both humoral and cell-mediated immunological responses are important in the control of IBV infections. Any current field challenge causes immuno-suppression which predisposes birds for IB. Innate immunity is crusial for the control of IBV field infections and its early activation is mainly due to action of gamma-interferon, as a result of the action of macrophages in addition to other sustances released during early inflammatory process (Kotani *et al.*, 2000).The adoptive responses of the birds against the virus are the basis of the development of humoral and cell-mediated immune responses. These responses are essential, as the presence of high titers of systemic antibodies in circulation is correlated positively even in the situation of failure in detecting IBV in kidneys and in genital tracts as well as the absense of decreased egg quantity and/or qualities (Mondale and Naqi, 2001).

The presence of Ig A (Immunoglobin A) in the respiratory mucosa is associated with resistance to infection, conferring its role in the inhibition of IBV infection as its preferred site. As the chicken can produce up to 100 mg/kg body weight/day of Ig A, as opposed of only 30 mg/kg body weight/day of IgG. The need and importance of such mucosal response may be inferred.

The generation of immune response against IBV is very complex. The susceptibility of birds is influenced by their MHC (major histocompatibility complex) genotype. MHC determines the quality of humoral and cell-mediated reponses, as the MHC molecules are responsible for binding the antigen epitope and to present it to T-lymphocytes (Jonier *et al.*, 2007). The presentation of main epitopes including the S1 spike protein by specialized antigen presenting cells (APCs) is the starting point of the generation of humoral (when the epitopes bind to MHC–II molecules) or cell-mediated (when the epitopes bind to MHC–I molecules) responces.

On the 6th day of embryonic life, the thymus begins to be colonized by undifferentiated cells that becomes T-lymphocytes, reaching to peak in 15–16 days. After 10th day of embryonic life, the bursa starts to be colonized. At hatch, these organs are anatomically developed but further functional maturation is accelerated by external stimuli. As environmental challenges are very different, the presence of maternally derived antibodies (MAbs) in newly hatched chicks during the early stages of life ensures a response to the external challenge. The importance of that passive protection is different according to the disease, but IgG antibodies transmitted from hens to chicks via egg yolk can usually be detected in the serum and respiratory mucosa of chicks.

Different studies have been shown that MAbs may overcome the challenges for an initial period that varies from 7-14 days (Mondal and Naqi, 2001). Sufficient Ig G levels may be present for at least 18 weeks of age, ensuring rapid reaction when a challenge emerges. Pei and Colisson (2003) pointed out that vaccinating 1 day old chick may be ineffective but may be an important time for a first controlled vaccine induced challenge.

Cytotoxic T-lymphocytes (CTLs) are detected 3 days after infection with IBV and reach to peak within 10 days. During this period, infection is normally resolved and the virus is eliminated (Pei and Colisson, 2005). After that, the number of CTLs drops together with the virus titers in the respiratory system and kidneys. After 2 weeks of infection, the number of CTLs doesn't remain sufficient to protect the birds.

The presence of virus of an attenuated vaccine in the respiratory system is intended to prevent further colonization of wild viruses. This phenamenon is called as "immune exclusion" and as above maintioned in the case of IBV, there may be undesirable consequences over time, not only due to the selection pressure but also by reversion and mutation of the vaccine virus, which may become uncontrolable variable in the medium run.

Neutralizing antibodies induced during infectious bronchitis are directed against the "S1 protein" of the surface spike. After infection, IgM peaks in 12 days and falls down upto day 21st post-infection. The presence of IgG is not normally detected until day 10 post-infection but it increases exponentially between day 20 and 30 post-infection. The induction of cytotoxic T-lymphocytes by CD8+ T-lymphocytes is very important in early stages of infection. These cells are detected for a long period at decreasing levels, as a result of a disease in challenge (Colisson *et al.,* 2000). The protective immunity against IBV is complex and doesnot result only from the bird's sensitization with the specific spike protein or from interractions of specific antibodies in a future challenge.

Vaccines and Vaccination

Purpose of Vaccination

1. To prevent the morbidity and mortality associated with respiratory disease caused by IBV.

2. In pullets, to protect the developing oviducts and kidneys due to damage resulting from infections with certain strains of IBV.

3. Prevention of egg production drops, poor egg shell quality and internal egg quality problems which can result due to infections of IBV in laying hens.

4. In breeder flocks, vaaccination promotes high levels of antibodies which is transferred passively as MAbs and provide only partial protection to the progeny chicks against early IBV cahllenges.

5. In broilers, to prevent morbidity, poor growth, condemnation at meat processing and mortality associated with IB outbreaks.

IB Vaccine Strains

IB vaccination programme requires multiple vaccinations. The first vaccination utilizes milder strains such as Massachusetts, H120 or Connecticut applied by a less invassive/reactive route of administration. Subsequent revaccinations may involve more immunogenic and more invassive/reactive strains. Secondary vaccinations broaden cross-protection and boost immunity stimulated by primary vaccination. Secondary vaccinations may be safely applied by more invassive/reactive routes of administration. Fine spray route is the most invassive/reactive route of administration followed by coarse spray and the least invassive/reactive route of administration is drinking water, intra-ocular or intranasal routes.

Commercial vaccines containing the same strain may differ considerably in immunogenecity and reactivity depending upon the laboratory manipulations used to develop the master seed. IB vaccine strain selection should be based upon the presence of field strains common in the area where it is to be used. New IB vaccine strains should not be introduced in to an area where they are not known to exist.

The Ark (Arkansas), D274, H120, Massachusetts and Connecticut strains are less invassive/reactive IB vacine strains and generally applied through intra-ocular, intranasal or in drinking water. Mass II, JMK, and D1466 strains are considered as moderately invassive/reactive vaccine strain applied through coarse spray route of administration. H52 and Australian-T strains are the most invassive/reactive vaccine strains which are applied through fine spray route of administration.

Some precautions may be adopted while vaccinating birds with IB vaccines. Post-vaccination reactions in birds can occur after vaccinaton with IB vaccines. These reactions are characterized by tracheal rales, snicking, conjunctivitis and increased susceptibility to E. coli infections and in some cases damage to the oviducts of growing pullets. Clinical signs may develop 3–5 days post-vaccination and if uncomplicated or mild, subside in next 3–5 days. The factors exacerbating the occurance of post vaccination reactions may be:

1. Administration of vaccine as a fine spray *i.e.* through the most invassive route.
2. Vaccination of Mycoplasma positive flocks.
3. Vaccination of birds that are immuno-compromised, morbid or with compromised respiratory system.
4. Using strong IB vaccine strain such as H52 as a primary vaccination.
5. Low levels of immunity resulted due to long intervals between successive live vaccinations.
6. Live vaccinations in multiple age laying birds.
7. Large amount of ammonia and dust in the environment.

Layer/Breeder Vaccination

Programme – 1

The objective of this vaccination programme is to protect the oviducts, kidneys and respiratry tract tissues of growing pullets and to maintain good mucosal (local) immunity of respiratory tracts of laying hens by using only live attenuated IB vaccines. In breeder, the level of MAbs (transferred from hens to progeny chicks) will generally be low and variable. So, MAbs in case of IB, only provides partial protection to chicks in early period of life.

Massachusetts, Connecticut and H120 strains are commonly used for 1st and 2nd IB vaccinations as these strains are widely prevalent in the field and relatively milder in reactivity nature. These live attenuated IB vaccines produce local as well as systemic immunity but preferably produce local mucosal immunity in the upper respiratory tract tissues of birds. Vaccine is administered generally through intranasal route or in drinking water by beak dipping up to nasal opennings. This method doesnot miss any bird of flock but it is laborious and costly. The interval between 1st and 2nd (1st booster) vaccinations should be 2–3 weeks because of the short duration of primary immune response. However, the secondary imune responses produced during subsequent vaccinations have a longer duration of immunity. This allows for longer intervals of 4–6 weeks between successive booster vaccinations.

IB vaccine strains which are too strong and reactive should not be used for primary vaccination. However, these strains such as Mass II, H52, some Ark (Arkansas) and JMK can be used as booster doses during revaccination. Commercial IB vaccines are often combined with Newcastle disease live attenuated vaccines due to similarities in vaccination timings and route of administration. Intra-ocular or intranasal route of administration is much advantagious as it allows individual birds to be vaccinated and uniform application dosage receivings compared to drinking water or spray administration.

Laying flocks should be vaccinated with live attenuated vaccines at 30 to 90 days intervals as booster vaccinations to maintain a high level of immunity during the egg production period and to minimize the risk of post-vaccination reactions. Generally, mild IB vaccine strains like Massachusetts strain are used for this purpose. Connecticut, H120 and D274 vaccine strains can also be used if prevalent in the field. So, field research for strain prevalence and protectotyping is needed for tailor made vaccines manufactured for the areas in question.

Live attenuated IB vaccines may result in increased pathogenicity (revert to virulence) when the live IB vaccine virus is repeatedly passed from a vaccinated bird to susceptible birds of the flock (if vaccines are administered in drinking water or spray in place of intra-occular or intranasal leaving some birds missed in getting the vaccine doses). Live attenuated vaccinations of laying flocks can result in transient drops in egg production and adversely affect the egg qualities. This is very common when there are long intervals between succcessive vactinations causing decreased flock immunity. Reducing the intervals between successive vaccinations may eleminate this effect on egg production. The monovalent (single fraction) Newcastle disease live vaccine and IB live vaccines should not be combined and administered as single vaccine because each of these viruses may interfere with the replication of the other. For this specifically formulated and tailor made vaccines from manufacturers should only be used because the vaccine virus titers of ND and IB in these specific products are properly adjusted to avoied interference of replication of viruses with each other.

Programme – 2

The objective of this vaccination programme is use of live attenuated and inactivated (killed) IB vaccines to protect oviducts, kidneys and respiratory tract tissues of growing pullets as well as to maintain good mucosal (local) and systemic immunity in layers and breeders against IBV infections.

This vaccination programme (live IB vaccins followed by inactivated IB vaccine boosters) is an alternative to the use of only live IB vaccines in layers and breeders. In laying hens, the inactivated (killed) IB vaccines provide a long duration systemic immunity. In breeders, this programme results in high and uniform MAbs level in progeny chicks. MAbs against IB provide only partial progeny protection against challenge and reduce the incidence of post-vaccination reactions from IB hatchery spray vaccinations.

Birds are allowed firstly for successful priming with live attenuated IB vaccine using generally Massachusetts mild strain. After it, booster live attenuated vaccinations should be given firstly at the intervals of firstly 2–3 weeks (between 1st and 2nd vaccinations) and later on 4–6 weeks (between 2nd and successive vaccinations). An injection of inactivated (killed) vaccine using killed Massachusetts, Mass II, Ark, D274 and D1466 vaccine (polyvalent) strains is given usually 4 weeks prior to the onset of egg production (at 12th week of age) to allow sufficient time for full development of immune response. Birds successfully primed with live attenuated IB vaccines respond better to the inactivated (killed) IB vaccination. A minimum interval of 4 weeks between the last live IB vaccination and the inactivated (killed) IB vaccination is required for an acceptable response but 6 weeks interval is prefered. For hens beings molted, a second inactivated (killed) IB vaccine can be given through injection during the molting period *i.e.* at the age of 42 – 44 weeks.

Advantages of Live IB Vaccination Programme

1. Good local immunity.
2. Enhanced cross – protection.
3. Impoved immunity in older laying flocks.
4. Mass administration of vaccine.

Disadvantages of Live IB Vaccination Programme

1. Requirement of live vaccinations of laying birds.
2. Potential for post-vaccination reactions resulting in egg production loss and decreased quality of eggs.
3. Chances of reversion of virulence.
4. Lower immunity during peak egg production.

Advantages of Live/Inactivated IB Vaccination Programme

1. Long duration of immunity.
2. No spread and no reversion of virulence.
3. Minimizes the need of live vaccinations of laying hens.
4. Improved immunity during peak egg production.
5. Polyvalent vaccine can deliver many antigens in a single administration.
6. Each bird is immunized.

Disadvantages of Live/Inactivated IB Vaccination Programme

1. Increased cost of vaccine administration.
2. Requirement of individual bird handling.
3. Potential of accidental injection to vaccine administrator.
4. Possible transient decrease in egg production during 2nd half of lay period.
5. The residue and tissue reaction from vaccination with oil – adjuvented inactivated IB vaccines may be detectable for many weeks post-vaccination.

Broiler Vaccination

Programme – 1

The objective of this vaccination programme is to protect broiler's respiratory tract tissues against early infections of IBV using mild live attenuated IB vaccine in hatchery.

MAbs in case of IB are only partially protective to progeny chicks. So, hatchery IB vaccination might be needed in high IBV challenge areas. Hatchery IB vaccination is generally administered as coarse spray (moderately invassive) route, which stimulates only local immunity in the mucosal tissues of upper respiratory tracts of the birds. The immunity from hatchery vaccination is of short duration (2-3 weeks) and only limited to the upper respiratory tracts of birds. The presence of MAbs against IBV does not interfere significantly with the development of local immune response. Revaccination (1st booster) with live attenuated IB vaccine is usually necessary during growing period of the birds. A third vaccination (2nd booster) of live attenuated IB vaccine may be needed in longer-lived broilers or in high IBV challenge areas. Generally, Massachusetts strain live attenuated IB vaccine is used for this vaccination. However, Ark, Florida, JMK, D274, D1466 and other regional strains containing live attenuated IB vaccine may be used in areas where there is a history of outbreaks caused by these strains.

Use of such vaccine that is not specifically designed and manufactured for 1 day old chicks, can cause severe post-vaccination reactions. Chicks lacking MAbs for IB may experience stronger post-vaccination reactions after application of hatchery coarse spray administered live attenuated IB vaccine.

Programme – 2

The objective of this vaccination programme is to protect respiratory tracts of slightly older broiler flocks by using mild live attenuated IB vaccines, when hatchery vaccination is not practiced.

In this programme, slightly older broilers are to be immunized using mild live attenuated IB vaccines. The advantage of vaccinating sligtly older broilers are that the MAbs interference is less likely to occur and the bird's immune system is more developed. Revaccination (1st booster) with a live attenuated IB vaccine is usually necessary to provide sufficient immunity throughout the growing period. A 3rd (2nd booster) vaccination with live attenuated IB vaccine is often needed in longer-lived broilers (roasters) or birds in high IBV challenge areas. Generally, Massachusetts strain mild live attenuated IB vaccine is administered. However, other strains like Connecticut, H120, Ark, JMK, Mass II and D274 may also be used, if there are sufficient evidences of their presence in the area in question.

REFERENCES

Alexander, D. J., Gough, R. E. and Pattison, M. (1978). A long-term study of the pathogenesis of infectionsof fowls with three strains of avian infectious bronchitis virus. *Res. Vet. Sci.*, 24, 228 – 233.

Britton, P. and Cavanagh, D. (2007). Avian coronavirus diseases and infectious bronchitis vaccine development. In *Coronaviruses: Molecular and Cellular Biology*. Thiel, V., ed, Caister Academic Press, Norfolk, U.K., 161 – 181.

Cavanagh, D. (2003). Severe acute respiratory syndrom vaccine development: Experiences of vaccination against avian infectious bronchitis coronavirus. *Avian Pathol.*, 32, 567 – 582.

Cavanagh, D. (2007a). Coronavirus avian infectious bronchitis virus. *Veterinary Research*, 38, 281 – 297.

Cavanagh, D. and Gelb, J., Jr. (2008). Infectious bronchitis. In: *Diseases of Poultry*, Twelfth Edition. Saif, Y. M., Fadly, A. M., Glisson, J. R., McDougald, L. R., Nolan, L. K. and Sweyne, D. E., Eds. Wiley-Blackwell, Ames, Iowa, USA, 117 – 135.

Canavagh, D., Mawditt, K., Welchman, D., De, B., Britton, P. and Gough, R. E. (2002). Coronaviruses from pheasants (Phasianus colchicus) are genetically closely related to coronaviruses of domestic fowl (infectious bronchitis virus) and turkeys. *Avian Pathol.*, 31, 81 – 93.

Colisson, E. W., Pei, J., Dzielawa, J. and Seo, S. H. (2000). Cytotoxic T-lymphocytes are critical in the control of infectious bronchitis virus in poultry. *Developmental and Comparative Immunology*, 24, 187 – 200.

Enjuanes, L., Almazan, F., Sola, I. and Zuniga, S. (2006). Biochemical aspects of coronavirus replication and virus-host interraction. *Annual Review Microbiology*, 60, 211 – 230.

Kotani, T., Shiraishi, Y., Tsukamoto, Y., Kuwamura, M., Yamae, J., Sakuma, S. and Gohda, M. (2000). Epithelial cell kinetics in the inflammatory process of chicken trachea infected with infectious bronchitis virus. *J. Vet. Med. Sci.*, 62(2), 129 – 134.

Lewicki, D. N. and Gallagher, T. M. (2002). Quatenary structure of coronavirus spikes in complex with carcinoembryonic antigen-related cell adhesion molecule cellular receptors. *The journal of Biological Chemistry*, 277(22), 19727 – 19734.

Mondal, S. P. and Naqi, S. A. (2001). Maternal antibody to infectious bronchitis virus: its role in protection against infection and development of active immunity to vaccine. *Veterinary Immuology and Immunopathology*, 79, 31 – 40.

Pei, J. and Colisson, E. W. (2003). Memory T-cells protect chicks from acute infectious bronchitis virus infection. *Virology*, 306, 376 – 384.

Pei, J. and Colisson, E. W. (2005). Specific antibody secreting cells from chickens can be detected by three days and memory B-cells by three weeks post-infection with the avian respiratory coronavirus. *Developmental and Comparative Immunlogy*, 29, 153 – 160.

Woo, P. C. Y., Lau, S. K. P., Huang, Y. and Yuen, K. Y. (2009). Coronavirus diversity phylogeny and interspecies jumping. *Experimental Biology and Medicine*, 234, 1117 – 1127.

Woo, P. C. Y., Lau, S. K. P., Lam, C. S. F., Lau, C. C. Y., Tsang, A. K. L., Lau, J. H. N., Bai, R., Teng, J. L. L., Tsang, C. C. C., Wang, M., Zeng, B. J., Chan, K. H. and Yuen,

K. W. (2012). Discovery of seve novel mammalian and avian coronaviruses in the genus Deltacoronavirus supports bat coronaviruses as the gene source of Alphacoronavirus and Betacoronavirus and avian coronaviruses as the gene source of Gammacoronavirus and Deltacoronavirus. *J. Virol.*, 86, 3995 – 4008.

Yu, L., Jiang, Y., Low, S., Wang, Z., Nam, S. J., Liu, W. and Kwang, J. (2001). Characterization of three infectious bronchitis virus isolates from China associated with proventriculus in vaccinated chickens. *Avian Dis.*, 45, 416 – 424.

Zeng, F., Hon, C. C., Yip, C. W., Law, K. M., Yeung, Y. S., Chan, K. H., Peiris, J. S. M. and Leung, F. C. C. (2006). Quantitative coparison of the efficiency of antibodies against S1 and S2 subunit of SARS coronavirus spike protein in virus neutralization and blocking of receptor binding: Implications for the functional roles of S2 subunit. *Federation of European Biochemical Societies Letters*, 580, 5612 – 5620.

Chapter 6

Infectious Laryngotracheitis

Infectious laryngotracheitis (ILT) is an infectious viral respiratory tract disease naturally and primarily occurring in chickens and resulting in to severe production losses in terms of mortality of infected broilers, pullets and adult birds, decreased weight gain and decreased egg production. ILT was firstly described in 1925 (May and Thittsler, 1925). It has been reported from many countries of the world in which it remains as a serious disease mainly in areas of intensive poultry production and large concentration of chickens such as North America, South America, Europe, China, south-East Asia and Australia. Chicken flocks in some regions of many countries where generally multiple aged birds are reared at production sites (industrial and backyard flocks), ILT remains present endemically. However, serious disease outbreaks continue to occur periodically whenever ILT virus (ILTV) strains can move from persistently infected flocks to non-vaccinated birds.

ILT disease has been listed with in "list-B" of the office International des Epizootics (OIE, 1999). So, close attention must be needed towards ILT infection or vaccination status of chickens when birds are being moved either nationally or internationally.

ILTV is a DNA virus and member of the family Herpesviridae, subfamily Alpha-Herpesvirinae, genus Varicellovirus and species is named as gallid Herpesvirus (Murphy *et al.*, 1995). ILTV is of icosahedral symmetry measuring 100 – 110 nm in diameter. Nucleocapsid contains 162 capsomeres which are hexagonal in cross-section. The core consists of a fibrillar pool on which the single molecule of dsDNA is wrapped. Virus is irregularly enveloped bearing glycoprotein spikes on outer surface (Murphy *et al.*, 1995). Five major proteins of envelop have been reported to be the major immunogens of ILTV. ILTV strains are considered to be antigenically homogenous. However, the use of polymerase chain reaction (PCR) and restricted fragment length polymorphism (RFLP) has enabled differentiation of field and attenuated vaccine type virus strains (Chang, *et al.*, 1997). ILTV is readily inactivated by low heat (60°C for 15 minutes), ether, chloroform and too extremes of pH (Bagust and Guy, 1997).

Epidemiology

Chicken is the only significant natural primary host species for ILTV and no other reservoir species has been recognized. However, sometimes pheasant and peafowl can be naturally infected through contact with chickens actively shedding ILTV (Guy and Bagust, 2003). The chief sources of ILTV are:

1. Clinically infected chickens.
2. Chickens which are latent carriers of infection.
3. Fomites, poultry farm equipments and personnel contaminated with ILTV.

ILT viruses being shed after the latent period are other source of viruses capable of causing disease in susceptible birds. ILT is a slowly spreading disease, so controllable and can be eradicated (Mallinson *et al.*, 1981). An outbreak can be successfully managed, if diagnosed early and followed by vaccination of unaffected birds which induce adequate immunological protection before they become exposed.

ILT vaccine virus has been shown to spread readily from vaccinated to non-vaccinated chickens (Andreasen *et al.*, 1989). Such spread should be avoided, as spread to non-vaccinated birds results in vivo passage and possible reversion of vaccine virus to virulence (Guy *et al.*, 1991). As, vaccination can produce latently infected carrier birds, it is only recommended for use in such geographical areas where the disease is endemic. Avoid the vaccination in the geographical areas from where this disease in not reported.

Pathogenesis

Chickens are infected by ILTV through upper respiratory and ocular routes (Beaudette, 1937). Ingestion may be another possible route of infection in association with nasal epithelial contact (Robertson and Egerton, 1981). ILTV infection of upper respiratory tract in susceptible chickens is followed by intense viral replication. ILTV initiates infection by attachment to cell receptors followed by fusion of the viral envelop with the host cell plasma membrane. The nucleocapsid is released in to the cytoplasm and transported to the nuclear membrane where viral DNA is released from the nucleocapsid and migrates in to the nucleus through nuclear pores. Replication of viral DNA occurs within the nucleus (Guo *et al.*, 1993).

Transcription of ILTV DNA occurs in a highly regulated, sequentially ordered cascade similar to that of other Alpha-Herpesviruses (Prideaux, *et al.*, 1992). Approximately 70 virus coded proteins are produced after translation process, out of which several are enzymes and DNA-binding proteins that regulate viral DNA replication. But most of them are viral structural proteins. ILT DNA replication occurs by rolling circle mechanism with the formation of concatemers which are cleaved in to monomer units and further packaged in to preformed nucleocapsids with in the nucleus. DNA filled nucleocapsids acquire envelops by migration through the inner lamellae of the nuclear membrane. Enveloped particles then migrates through the endoplasmic reticulum and be accumulated with in vacuoles in the cytoplasm (Guo *et al.*, 1993). Enveloped viruses are released by cell lyses or by vacuolar membrane fusion followed by exocytosis.

Thus, ILTV usually remains present in tracheal tissues and its secretions for 6–8 days post-infection (Bagust *et al.*, 1986). The virus may remain at very low levels up to 10 days post-infection (Williams *et al.*, 1992). No clear evidence exists for a viremic phase of infections. So, local and cell mediated immunity is of greater concern in this ILTV infections.

Thus, the main target organ system for ILTV infection and disease is the respiratory tract. The epithelium of trachea and larynx (in birds it is called as "SYRINX", so ILT is probably a misnomer) is always infected whilst other mucus membranes such as the conjunctiva, respiratory sinuses, air sacs and lung tissues may also be infected. Whether chickens are exposed to ILTV by nasal, oral, conjunctival (ocular) or sinus routes, the most active replication of ILTV occurs with in tissues of trachea and larynx (syrinx). Active viral replication occurs only during the 1st week after infection, although low levels of sporadic infections may be detected up to 10 days post-infection (Bagust *et al.*, 1986). From 10 days to approximately 4 weeks after tracheal infection, shedding of ILTV may be ceased. A latent phase of infection may be established through ILTV invasion of nervous tissues. Invasion of Trigeminal ganglion by ILTV has been found to occur regularly during acute phase (3-6 DPI) of ILTV infections by field and vaccine strains (Bagust *et al.*, 1986). The exact route of infection of Trigeminal nerve ganglion is not known but neural migration is strongly inferred as this ganglion provides the main sensory innervations to the tissues of upper respiratory tract, the mouth and the eyes whilst the distal ganglia are also involved in the sensory innervations of the trachea. It is now confirmed that the tracheal ganglion is the main site of latency of ILTV.

Onset of the latent phase of ILTV infection commences from the immediate post-acute phase of infection *i.e.* 7-10 days after tracheal exposure. Latent ILTV infections are not readily demonstrable during the first few months after infection (Hughes *et al.*, 1991) probably reflecting initial high level of host immune control. Subsequently, sporadic reactivation of latent ILTV infection with shedding of low level of infectious viruses in to the trachea is maintained throughout the life.

The rates of shedding of ILTV in to trachea can be significantly increased by the stress of either onset of lay or mixing with unfamiliar birds. In this case, the latently infected chicken can act as an unsuspected reservoir hosts and further enable ILTV to infect susceptible chickens (Hughes *et al.*, 1989). It should be understood that establishment of latency by ILTV is the biological survival mechanism which enables ILTV to evade host's immune surveillance and to persist even in small flocks of chickens over generations.

Clinical Signs

Clinical signs generally appear 6-12 days following natural exposure (Kernohan, 1931a). Clinical signs which are characteristics of ILT include nasal discharges, moist rales, coughing and gasping (Beach, 1926). Two forms of ILT disease are recognized:

1. Severe epizootic form
2. Mild enzootic form

1. Severe Epizootic Form

This form is characterized by marked dyspnoea, gasping, coughing, expectoration of blood stained mucus and high mortality (Beach, 1926). This form causes high morbidity (90– 100 per cent) and variable mortality generally between 5–70 per cent and averages 10 – 20 per cent (Hinshaw *et al.*, 1931).

Gross lesions may be found in conjunctiva and throughout the respiratory tract of ILTV infected chickens but are more consistently observed in trachea and larynx (syrinx). The changes in tracheal and laryngeal tissues may be mild consisting only the excess mucus (Linares *et al.*, 1994) or severe with hemorrhagic and/or diphtheritic changes. Mucoid inflammation may be observed. Diphtheritic changes are common and may be seen as mucoid casts that extend the entire length of trachea. In other cases, severe hemorrhages in to tracheal lumen filled with blood casts or blood mixed mucus and necrotic tissues. Inflammation may extend down to the bronchi, lungs and air sacs.

Early microscopic changes in tracheal mucosa include the loss of goblet cells and infiltration of the mucosa with inflammatory cells. As the viral infection progresses, cells enlarge, become decilliated and become edematous. Multinucleated cells (syncitia) are formed. Lymphocytes, histiocytes and plasma cells migrate in to the mucosa and sub mucosa. Later, cell destruction and desquamation result in a mucosal surface either covered by thin layers of basal cells or lacking any epithelial covering. Hemorrhages may occur in cases of severe epithelial destruction and desquamation due to rupture of blood capillaries (Guy, *et al.*, 1990). Eosinophilic intranuclear inclusion bodies may be found in epithelial cells of trachea, larynx (syrinx) and conjunctiva. These inclusion bodies are generally present only in early stages of ILT infection.

2. Mild Enzootic Form

Clinical signs associated with this form of disease include unthriftiness, decreased egg production, watery eyes, conjunctivitis, swelling of infra orbital sinuses, nasal discharge and hemorrhagic conjunctivitis. During recent years, this form has become more common particularly in more intensive poultry production areas. This form results in low morbidity (5 per cent) and very low mortality *i.e.* 0.1 – 2 per cent (Linares *et al.*, 1994).

Gross lesions may consist of conjunctivitis, sinusitis and mucoid tracheitis (Linares *et al.*, 1994). Sometimes edema and congestion of epithelium of conjunctiva and infra-orbital sinuses may be the only gross lesions observed in this mild form of ILT.

Microscopic lesions consist of inflammatory cells infiltration. Eosinophilic intranuclear inclusion bodies are seen in severe epizootic form of ILT especially in early stages of infection. Regeneration commences approximately after 6 days post-infection. Intranuclear inclusion bodies may not be visible during and after regeneration period. Thus, the appearance of these inclusion bodies is transient and inability of their presence doesn't exclude the diagnosis of ILT.

Diagnosis

Diagnosis is primarily based on clinical signs and lesions but differentiation from other respiratory diseases like ND (Newcastle disease) and IB (Infectious bronchitis) needs laboratory assistance, which may be:

1. Histological examination of trachea.
2. Detection of ILTV.
3. Detection of specific antibodies.

Histological Examination of Trachea

The development of eocinophilic intranuclear inclusion bodies in the respiratory tract and conjunctival epithelium is pathognomnic for ILT. Epithelial hyperplasia gives rise to multinucleated cells (syncitia) in which these intranuclear inclusion bodies may be apparent. Hemorrhages in lamina propria of trachea may also be present.

Detection of Virus (ILTV)

Isolation of virus is promisingly diagnostic in embryonated chicken eggs or cell cultures. Exudates and epithelial scrapings from trachea is collected in swabs and are emulsified in nutrient broth. Supernatants of the broth cultures are inoculated on to the dropped CAM (chorio-allantoic membrane) of 10-12 days old fertile chicken eggs or in to preformed monolayers of suitable cell cultures. Samples of tissues or swabs should be taken as soon as possible after the onset of clinical signs, since isolation attempts may be unsuccessful beyond 6-7 days post-infection (Bagust and Guy, 1997). Inoculation of embryonated eggs with ILTV results in the production of "POCKS" on the CAM and inoculation of cell culture causes "syncitium formation".

If there is co-contamination of other viruses, PCR is most sensitive molecular method of diagnosis compared to other methods for diagnosis of ILTV (Williams *et al.*, 1994).

Detection of Antibodies

Various techniques and methods have been used to detect antibodies in serum of chickens infected with ILTV. AGID (agar gel immuno diffusion) is very simple method using hyper immune serum raised against ILTV (Jordan and Chubb, 1962). AGID may be valuable while differentiating ILTV from diphtheritic form of fowl pox. Immuno-florescence (IF), immune-peroxidase (IP) test and antigen capture ELISA are novel immunological methods of ILTV diagnosis (Adair, 1985).

Treatment

There is no specific treatment for ILT. Secondary bacterial infections should be treated using suitable antibiotics along with supportive therapies.

Prevention and Control

It includes effective biosecurity and vaccination of birds. The effective biosecurity measures may be avoiding exposing of susceptible chickens via contaminated fomites, site quarantine and hygiene in preventing the movement of potentially contaminated personnel, feed, equipments and birds are key points in prevention of ILT (Kingsbury and Jungherr, 1958). Raising multi aged chickens at same or nearby site should not be practiced.

For control of an ILT outbreak, the most effective approach is co-ordinated effort to obtain a rapid diagnosis, institute vaccination programme and prevent further virus spread (Bagust, 1992). Spread of ILTV between sites can be prevented by adopting appropriate biosecurity measures. The ILTV is readily inactivated outside the host chicken by disinfectants and low level of heat, thus spread of disease between successive flocks housed in the same building can be prevented by adequate cleaning precautions.

Immunological Considerations

A variety of immune responses are generated by the immune system of chickens following infection of ILTV (Jordan, 1981; Hitchner, 1975). Best known are the virus neutralizing antibodies which become detectable in the serum within 5–7 days of tracheal exposure, peak around 21 days and exist over the next several months at low level which can persist for a year or more. Mucosal antibodies activity (Ig G and Ig A) which are capable of binding to ILTV antigen and low level of virus neutralizing (VN) antibodies activity become detectable in tracheal secretions and washings from infected birds approximately 7 days post-infection (York *et al.*, 1989) and plateau at 10 to 28 days post-infection. However, numerous laboratory and field studies have independently confirmed that immune protection to ILTV challenge is neither indicated nor conferrable by the presence of serum antibodies or MAbs (Bagust and Guy, 1997).

Cell-mediated immunity is known to be the main protective immune response in ILTV infection and in vaccination (Jordan, 1981). Studies using vaccinated-bursectomized chickens have demonstrated that even tracheal mucosal antibodies are not essential in preventing the replication of virus in vaccinated chickens (Fahey and York, 1990). Rather, the effective mechanism of protection against ILTV infection is likely to be local cell-mediated immune response in the trachea.

Primary vaccination with live attenuated ILT vaccine strains assures only partial protection against challenge up to 3–4 days post-vaccination and complete protection against challenge after 1 week post-vaccination (Hitchner, 1975, Jordan, 1981). High level of protection occurs between 15–20 weeks post-vaccination. Revaccination with live ILT vaccine may or may not assist in maintaining protection level (Jordan, 1981).

Vaccines and Vaccination

The objective is to (1). Prevent morbidity, mortality and upper respiratory tract disease associated with clinical outbreaks of ILT in commercial layers, breeders and broilers. (2). To prevent drops in egg production that may occur from ILT infections in commercial layers and breeders. (3). To stimulate immunity in susceptible birds at high risk of infection during clinical outbreaks of ILT.

Layer/Breeder Vaccination

Live attenuated vaccine is generally used to immunize chickens against ILTV infections. This live vaccine is only for use in commercial layers and breeders in the areas where ILT outbreaks have occurred.

Two successive live vaccinations are usually needed for establishment of sufficient flock immunity. The 1st vaccination with live attenuated ILT vaccine is practiced as early as 10 days of age to provide early flock protection. The 2nd vaccination again with live attenuated ILT vaccine is administered at least 4 weeks prior to the onset of egg production *i.e.* at the age of 12th week as generally after 16th week of age egg laying is started in well managed flocks. Intra-ocular route of vaccine administration is usually followed for 1st vaccination as no bird is missed in this way. Vaccination of susceptible birds during an ILT outbreak is adopted because of the long incubation period of ILTV and slow spread of ILT disease.

All the birds on a farm should be vaccinated within a possible shortest period of time. If ILT vaccine is given as spray or in drinking water, post-vaccination reactions may occur which are characterized by mild clinical signs upper respiratory tract and conjunctivitis. The vaccination of Mycoplasma infected birds should not be done. Such birds also should not be vaccinated, which are having any respiratory disease. No other vaccine should be administered within 10 days of vaccination with ILT vaccine. As vaccinated birds can assumed to be latently infected by vaccine viruses and shed ILT viruses many weeks or months post-vaccination. For this reason, the vaccinated birds should not be mixed with unvaccinated birds or transported in to such areas where ILT vaccination is not being practiced.

Broiler Vaccination

This live vaccination programme is practiced for broiler flocks in only those areas having history of ILT outbreaks.

Broilers are usually not vaccinated for ILT disease except during outbreaks in the area in question. Generally, one live ILT vaccination is sufficient for broilers and should be given around 10 days of age by intra-ocular route preferably. Drinking water route is not followed unless advised by the vaccine manufacturers for their tailor made vaccines. Single day vaccination is adopted for all the birds at farm. Post-vaccination reactions may occur sometimes as in layer and breeder flocks.

REFERENCES

Adair, B. M., Todd, D., McKillop, E. R. and Burns, K. (1985). Comparison of serological tests for detection of antibodies to infectious laryngotracheitis virus. *Avian Pathol.*, 14: 461-469.

Andreasen, J. R. Jr., Glisson, J. R., Goodwin, M. A., Resurreccion, R. S., Villegas, P. and Brown, J. (1989). Studies of infectious laryngotracheitis vaccines: Immunity in layers. *Avian Diseases,* 33: 524-530.

Bagust, T.J. (1992). Laryngotracheitis. In: *Veterinary Diagnostic Virology: A Practitioner's Guide.* Mosby Year Book, St Louis, Missouri, 40-43.

Bagust, T. J., Calnek, B. W. and Fahey, K. J. (1986). Gallid-1 herpesvirus infection in the chicken. 3. Reinvestigation of the pathogenesis of infectious laryngotracheitis in acute and early post-acute respiratory disease. *Avian Dis.,* 30: 179-190.

Bagust, T. J. and Guy, J. S. (1997). Laryngotracheitis. In *Diseases of Poultry,* 10th Ed. (B.W. Calnek with H. J. Barnes, C. W. Beard, L. R. McDougald and Y. M. Saif, eds). Iowa State University Press, Ames, 527-539.

Beach, J. R. (1926). Infectious bronchitis of fowls. *J. Am. Vet. med. Assoc.,* 68: 570-580.

Beaudette, F. R. (1937). Infectious laryngotracheitis. *Poultry Science,* 16: 103-105.

Chang, P. C., Lee, Y. L., Shien, J. H. and Shieh, H. K. (1997). Rapid differentiation of vaccine strains and field isolates of infectious laryngotracheitis virus by restriction fragment length polymorphism of PCR products. *J. Virol. Meth.,* 66 (2): 179-186.

Fahey, K. J. and York, J. J. (1990). The role of mucosal antibody in immunity to infectious laryngotracheitis virus in chickens. *J. Gen. Virol.,* 71: 2401-2405.

Guo, P., Scholz, E., Turek, J., Nordgreen, R., and Maloney, B. (1993). Assembly pathway of avian infectious laryngotracheitis virus. *American Journal of Veterinary Research,* 54: 2031-2039.

Guy, J. S. and Bagust, T. J. (2003). Laringotracheitis. In *Diseases of Poultry,* 11th Ed. (Y.M. Saif with H. J. Barnes, A. M. Fadly, J. R.Glisson, L. R. McDougald and D. E. Swayne, eds). Iowa State University Press, Ames: 121-134.

Guy, J. S., Barnes, H. J. and Smith, L. G. (1990). Virulence of infectious laryngotracheitis viruses: comparison of modified-live vaccine viruses and North Carolina field isolates. *Avian Dis.,* 34: 106-113.

Guy, J. S., Barnes, H. J. and Smith, L. G. (1991). Increased virulence of modified-live infectious laryngotracheitis vaccine virus following bird-to-bird passage. *Avian Diseases,* 35: 348-355.

Hinshaw, W. R., Jones, E. C. and Graybill, H. W. (1931). A study of mortality and egg production in flocks affected with laryngotracheitis. *Poultry Science,* 10: 375-382.

Hitchner, S.B. (1975). Infectious laryngotracheitis: the virus and the immune response. *Am. J. Vet. Res.,* 36: 518-519.

Hughes, C. S., Gaskell, R. M., Jones, R. C., Bradbury, J. M. and Jordan, F. T. W. (1989). Effects of certain stress factors on the re-excretion of infectious laryngotracheitis virus from latently infected carrier birds. *Res. Vet. Sci.,* 46: 247-276.

Hughes, C. S., Williams, R. A., Gaskell, R. M., Jordan, F. T. W., Bradbury, J. M., Bennett, M. and Jones, R. C. (1991). Latency and reactivation of infectious laryngotracheitis vaccine virus. *Arch. Virol.,* 121: 213-218.

Jordan, F. T. W. (1981). Immunity to infectious laryngotracheitis. In: *Avian Immunology* (M.E. Ross, L.N. Payne and B.M. Freeman, eds). British Poultry Science Ltd, Edinburgh, 245-254.

Jordan, F. T. W. and Chubb, R. C. (1962). The agar gel diffusion technique in the diagnosis of infectious laryngotracheitis (ILT) and its differentiation from fowl pox. *Res. Vet. Sci.*, 3: 245-255.

Kernohan, G. (1931a). Infectious laryngotracheitis in fowls. *Journal of American Veterinary Medical Association*, 78: 196-202.

Kingsbury, F. W. and Jungherr, E. L. (1958). Indirect transmission of infectious laryngotracheitis in chickens. *Avian Diseases*, 2: 54-63.

Linares, J. A., Bickford, A. A., Cooper, G. L., Charlton, B. R. and Woolcock, P. R. (1994). An outbreak of infectious laryngotracheitis in California broilers. *Avian Dis.*, 38: 188-192.

Mallinson, E. T., Miller, K. F. and Murphy, C. D. (1981). Co-operative control of infectious laryngotracheitis. *Avian Dis.*, 25: 723-729.

May, H. G. and Thittsler, R. P. (1925). Tracheo-laryngotracheitis in poultry. *Journal of American Veterinary Medical Association*, 67: 229-231.

Murphy, F. A., Fauquet, C. M., Bishop, D. H. L., Ghabrial, S. A., Jarvis, A. W., Martelli, G. P., Mayo, M. A. and Summers, M. D. (eds) (1995). *Virus Taxonomy: Classification and Nomenclature of Viruses*. Sixth report of the International Committee on Taxonomy of Viruses. Springer-Verlag, Vienna and New York, 586 pp.

Office International des Epizooties (OIE) (1999). International animal health code: mammals, birds and bees, 8th Ed. OIE, Paris, 468 pp.

Prideaux, C. T., Kongsuwan, K., Johnson, M. A., Sheppard, M. and Fahey, K. J. (1992). Infectious laryngotracheitis virus growth, DNA replication, and protein synthesis. *Archives of Virology*, 123: 181-192.

Robertson, G. M. and Egerton, J. R. (1981). Replication of infectious laryngotracheitis virus in chickens following vaccination. *Australian Veterinary Journal*, 57: 119-123.

Williams, R. A., Bennett, M., Bradbury, J. M., Gaskell, R. M., Jones, R. C. and Jordan, F. T. W. (1992). Demonstration of sites of latency of infectious laringotracheitis virus using the polymerase chain reaction. *Journal of General Virology*, 73: 2415-2430.

Williams, R. A., Savage, C. E. and Jones, R. C. (1994). A comparison of direct electron microscopy, virus isolation and a DNA amplification method for the detection of avian infectious laryngotracheitis vims in field material. *Avian Pathol.*, 23: 709-720.

York, J. J., Young, J. G. and Fahey, K. J. (1989). The appearance of viral antigen and antibody in the trachea of naïve and vaccinated chickens infected with infectious laryngotracheitis virus. *Avian Pathol.*, 18: 643-658.

Chapter 7

Fowl Pox Disease

Fowl Pox is an infectious disease of chickens and turkeys, caused by a DNA virus of the genus Avipox, subfamily Chordopoxvirinae of the family Poxviridae (Tripathy, 1993). Its distribution is world wide. It is slow spreading disease and characterized by the formation of proliferative lesions and scabs on the skin and diphtheritic lesions in the upper part of the digestive and respiratory tracts. In the case of cutaneous form, the mortality rate is usually low and the affected birds are more likely to recover than those with in the diphtheritic form. In the dephtheritic form, proliferative lesions involving the nasal passage, larynx (syrinx) and trachea can result in respiratory distress leading to death because of suffocation. Fowl pox causes a transient drop in egg production in laying chickens and a reduced growth rate in younger birds.

The genus Avipox virus is one of the eight genera of subfamily Chordopoxvirinae, all of which infect vertibrates. Viruses of other seven genera except Avipox virus infect mammals, while Avipox virus infects only birds. So, Avipox virus is not of zoonotic potential (Jarmin *et al.*, 2006).

Fowl pox virus (FPV) is the best studied and prototype species of the Avipox viruses. Avipox virus infections have been observed in more than 230 of the known 900 species of birds, spanning 23 orders. The mature virus (elementry body) is large, oval, double stranded DNA containing, break shaped and measures about 330x280x200 nm. The outer coat is composed of randomly arranged of surface tubules. The virus consists of an electron dense, centrally located biconcave core or nucleoid with two lateral bodies in each concavity and surrounded by an envelop. Envelop is developed in the cytoplasm of infected epithelial cells during maturation after replication (Jordan *et al.*, 1996). The 288 kilo base pair fowl pox virus genome encodes for over 250 genes.

Three most common strains of avipox virus have been identified which are fowl pox virus (affecting fowl), pigeon pox virus (affecting pigeons) and canary pox virus (affecting canaries). Turkey and quail pox viruses are also documented affecting turkeyes and quails respectively (Mandal and Johri, 2004). The strains vary in their virulence and have the ability to infect other avian species (Jarmin *et al.*, 2006).

Epidemiology

Fowl pox virus (FPV) can remain alive for 4–10 years in contaminated environment. Mosquitoes and other blood sucking insects can transmit the virus. Culex pipiens and Aedes aegypti mosquitoes are common mechanical vectors or transmitters of this virus from infected to susceptable hosts. The lesions are developed within 5–10 days after the infected vector mosquito was allowed to feed on a susceptable chicken. Mosquitoes that feed on infected birds play the most important role for both disease transmission and long term disease maintenance. However, the disease tends to be seasonal and occurs mainly after mosquito breeding seasons. Avian pox is transmitted when a mosquito feeds on infected bird that has viraemia (pox virus circulating in its blood) or when a mosquito feeds on virus-laden secretions seeping from a pox lesion and then feeds on another bird that is susceptible to that strain of pox virus. Fleas also may act as a mechanical vector for transmission of pox virus (Smits *et al.*, 2003).

On exposure to air-borne particles contaminated with pox virus can also result in infections when the virus enters the host body through abraded skin, conjuctiva or the mucus membrane lining that covers the front part of the eye ball and inner surfaces of the eyelids. Avian pox virus is unable to penetrate the unbroken skin but small abrasions are sufficient to permit infection (Tripathy, 1986). Infection through respiratory tract by inhalation is also possible (Pattison *et al.*, 2008). Individual handling of the birds at the time of vaccination may carry the virus on their hands and clothes and may unknowingly deposit the virus in the eyes or abrated skin of susceptible birds (Vegad, 2008). As new bird's species are being affected during recent years, fowl pox disease is becoming an emerging viral disease. The morbidity in fowl pox disease is 10–95 per cent. However, mortality ranges between 0–50 per cent (Tripathy, 1986).

The avian pox virus is resistant outside the host remaining viable in dry scales for long periods but sensitive to heat, lipid, phenol and other disinfectants.

Pathogenesis

Virus enters a skin cell and then spreads from cell to cell locally. Some viruses enter the blood to cause viraemia. Although, there is spread to internal organs through blood, no changes are usually seen. However, it is likely that there is some viral growth in certain internal organs like liver and spleen which leads to the development of secondary viraemia allowing the virus to enter again in to skin cells resulting in to a generalized disease.

Avian pox disease is considered to be a significant contributory factor to the cause of death for all birds examined post-mortem. The large peri-ocular skin lesions result in severely compromized vision and would have been vulnerable to predator attack. So, injuries from predator attacks are likely the ultimate cause of death. Spleenomegaly may also disposes the birds for predation (George, 2003).

The most common form of the disease is cutaneous form which consists of warty nodules that develop on the featherless parts of the body. This form of disease is usually self-limiting, the lesions regress and leave minor scars. However, the nodules can be enlarged and clustered thus resulting in to sight, breathing and feeding impairments. So, feed consumption and production are decreased. This form of disease is also called as "dry form" of disease.

The "wet form" of disease which is also called as "diphtheritic form" results in severe clinical signs causing interference with eating and/or breathing and ultimately leading to death of birds due to asphyxiation when trachea is affected (Rocke *et al.*, 2005).

Signs and Symptoms

Clinical signs observed in avian pox disease are weakness, emaciation, difficulty in swallowing, vision problems, reduction in egg production, soiled facial feathers, conjunctivitis, edema of eyelids, presence of characteristic wart-like growths on the unfeathered portions of the skin and/or formation of a diphtheritic membrane on the upper portion of digestive and respiratory tracts. Avian pox disease can occur in two forms: 1. Cutaneous or dry form and 2. Diphtheritic or wet form (Jordan *et al.*, 1996).

1. Cutaneous or Dry Form

This form of avian pox disease is characterized by cutaneous eruptions or wart-like nodules on the unfeathered body parts of fowl *viz.* comb, wattle, eyelid, feet, cloacal aperture and under the wings. In young chicks, these warty nodules are found in the corners of mouth, nostrils and eyelids. These are due to local epithelial hyperplasia. Removal of warts results in to bleeding. Firstly, the nodules appear as small whitis foci, which rapidly increase in size and become yellowish in colour as they develop. In some cases, closely adjoining lesions may coalesce and form large lesions which may be rough, gray or dark-brown in colour. After two weeks of development, the lesions may show the area of inflammation at their base and become hemorrhagic. The lesions then undergo a process of desication resulting in to scabe formation which may persist for further 1–2 weeks. If in the mean time, desicated scabs are removed, moist sero-purulent exudates are found underneath covering a bleeding and granulating surface. When the scab drops off a smooth scar may be left (Samour, 2004).

2. Diphtheritic or Wet Form

This form of disease is less common than dry form. Here, the eruptions on the mucus membranes are white, opaque and in the form of slightly elevated nodules. These nodules rapidly increase in size, often coalescing to become a yellowish, cheezy and necrotic material with the appearance of a pseudo-membrane. When these pseudo-membranes are removed, bleeding and erosions are resulted. The secondary invasion of bacteria aggravates the diphtheritic form of disease. The inflammatory prosess may extend from the mouth region to the sinuses particularly the infra-orbital sinuses resulting in tumor like swellings which may extend in to pharynx (sometimes up to trachea) causing respiratory disturbances. This form of disease is primarily a problem of young chickens and turkeys (Vegad, 2008).

Lesions

A mild form of fowl pox disease may remain unnoticed with only small focal lesions usually on combs and wattles. In severe forms of the disease, generalized lesions may occur on any part of the body such as comb, wattle, corners of mouth, around the eyelids, angle of beak, ventral surface of wings, legs and vent. Coalescence of the lesions around the eyelids can cause partial or complete closer of one or both eyes resulting in to vulnerability to predators because of impaired vision (Riper and Forrester, 2006).

In cutaneous/dry form of the disease, initially a small white, pink or yellow vesicle (blister) on unfeathered parts of the skin (feet, legs, base of the beak, eye margins and head) is formed. The vesicle is a result of the separation of the surface layer of the skin with the formation of the pockets of watery fluid rich in multiplying virus. The vesicles become nodules as they increase in size, coalesce and burst. Lymph from these cells congeals and scabes are formed. The surface of the nodules become rough and dry. Colour changes to dark-brown or black. Size and number of nodules depends upon the severity of the disease. More the severity of the disease, more the number and larger the size of nodules will be seen. Secondary bacterial infections may result in purulent discharges from these lesions. Later the scab falls off and a scar is formed on the site. It takes 2–4 weeks for complete healing (Vegad, 2008).

The diphtheritic/wet form involves the mouth, throat, trachea and lungs consisting of yellow to white, moderatly raised, moist cheese-like necrotic areas. A diphtheritic membrane is formed which may restrict respiratory air intake resulting in to laboured breathing, gasping and possible suffocation leading to death of infected birds. Lesions in the mouth, tongue and oesophagus interfere with the feeding and lesions of trachea often results in formaion of tracheal plugs leading to severe dyspnoea. This form of disease may resumbles with I.L.T. (Tripathy, 1986). This form is most likely to occur in wild birds. This form results in to greater morbidity an mortality. Spleenomegaly may be observed possibly in response to secondary bacterial infections (Winterfield and Reed, 1985).

Microscopic lesions consist of eosinophilic intracytoplasmic inclusion bodies (Bolinger bodies), present in the infected skin and respiratory tract mucosal cells. In dephtheritic form of disease, nodular hyperplasia (increased number of cells) of the mucosa is observed. Severe hyperplasia with ballooning of epidermal cells, multiple coalescing foci of necrosis and eosinophilic intracytoplasmic inclusion bodies (in ILT – eosinoplilic intranuclear inclusion bodies of DNA virus) are characteristically and pathognomonically present. On examination of the central creamy yellow coloured core of each lesion, consisted amorphous and acellular proteinaceous material is found. In addition to the Bolinger bodies, viral factories or elementary bodies which are called "Borrel bodies" remain present in cytoplasm of infected cells and considered as viral replication sites (Pattison *et al.,* 2008).

Diagnosis

Diagnosis is based on clinical signs and lesions present in the infected birds. However, confirmatory diagnosis is made by assistance of various laboratory methods.

Identification of FPV

FPV multiplies in cytoplasm of epithelial cells with the formation of large Bolinger bodies that contain smaller elementary bodies (Borrel bodies). These inclusion bodies can be seen in sections of cutaneous and diphtheritic lesions by the use of Hematoxylin and Eosin (H and E) stain or Giemsa stain (Tripathy *et al.*, 1973). The Borrel bodies can be detected in smears from lesions by Gimenez method (Tripathy and Hanson, 1976).

Smear Technique for FPV Detection

1. Place a drop of distilled water and some material from the lesion (cutaneous or diphtheritic) on a clean slide.
2. Prepare a thin smear by pressing the lesion with another clean slide and rotating the upper slide several times.
3. Air dry the slide and gently fix the smear over a flame.
4. Stain the smear for 5–10 minutes with freshly prepared stain (8 ml stock solution* of basic fuchsin mixed with 10 ml of phosphate buffer** (pH=7.5) and filtered through Whattman filter paper No.1).

*Stock Solution

A solution of basic fuchsin (5 gm) in 95 per cent ethenol (100 ml) is slowly added to a second solution of crystalline phenol (10 g) in distilled water (900 ml). This stock solution, kept in a tightly screw-capped glass bottle, is incubated for 48 hours at 37°C temperature and then stored at room temperature.

**Phosphate Buffer

pH = 7.5, $NaH_2PO_4.H_2O$ (2.47 g) and Na_2HPO_4 (11.65 gm) are added to distilled water (1000 ml) and stored at 4°C.

5. Wash the slide thoroughly with tap water.
6. Counterstain with Malachite green (0.8 per cent in distilled water) for 30 – 60 seconds.
7. Wash the smear with tap water and allow to air dry.
8. Examin the smear under oil-emersion (100x objective lense). The elementary bodies (Borrel bodies) appear red and are approximately 0.2–0.3 micrometer in size.

Isolation of FPV

FPV can be isolated by the inoculation of suspected material in to embryonated chicken eggs. Approximately 0.1 ml of tissue suspension of skin or diphtheritic lesions with the appropriate concentration of antibiotics, is inoculated on to the chorio-allantoic membranes (CAMs) of 9–12 days old developing chicken embryos. These eggs are incubated at 37°C temperature for 5–7 days and then examined for local white "pock lesions" or generalized thickening of CAMs. Histopathological examination of the CAM lesions will reveal eosinophilic intracytoplasmic inclusion bodies (Bollinger bodies) following staining with H and E stain (Tripathy and Reed, 1998).

Inoculation of the material infected with FPV in to cell cultures (primary chicken embryo fibroblasts, chicken embryo kidney cells, chicken embryo dermis cells and/ or permanent Quail cell line (QT–35) reveals "plaque formations" (Ghildyal *et al.,* 1989).

Molecular diagnostic methods like RFLP (Schnitzlein *et al.,* 1988) and PCR (Fallavena *et al.,* 2002) are also very useful.

Serological Tests

VNT (virus neutralization test), AGID (agar gel immuno difusion), passive HA (haemagglutination), FAT (fluoroscent antibody test), IP (immuno-peroxidase), ELISA and immunoblotting are very useful.

Treatment

No specific treatment is available.

Prevention and Control

Managemental practices should be implemented in poultry production areas to prevent the incidence of fowl pox disease by improving the hygiene. Vector transmission should be checked by controlling mosquitoes. A well programmed vaccination should be practiced to develop protective immunity in chickens.

Vaccination

The purposes of vaccination are:

1. To prevent the poor growth and mortality resulting from outbreaks of fowl pox disease in broilers, commercial layers and breeders.
2. To prevent egg production losses in laying birds.
3. To immunize birds at risk of infection during a clinical outbreak of fowl pox disease.

Layers/Breeders Vaccination

The objective of this vaccination is practicing fowl pox vaccination programme in layer and breeder flocks raised in areas with history of avian pox diaease outbreaks.

In areas of low challenge or with history of avian pox disease outbreaks occuring only in older birds, a single pox vaccination through wing web stabbing method may be sufficient. The vaccine will be of live attenuated type. In areas with a high challenge, prevalence of blood sucking insects or history of avian pox disease outbreaks in young birds, two live attenuated vaccinations through wing web stabbing method are recommended for flock protection.

Birds less than 6 weeks of age should be vaccinated with highly attenuated fowl pox or pigeon pox vaccine strains. Revaccination with fowl pox or turkey pox live vaccine is adopted. Vaccination of birds older than 6 weeks of age can be accomplished with less attenuated fowl pox vaccine strains. These strains are more immunogenic. Revaccination (1st booster dose) of the pullets should be practiced at least 4 weeks prior to the onset of egg production *i.e.* at 12th week of age by wing web stabbing method. Vaccination of the birds at risk of exposure during avian pox disease outbreaks is advisable due to the long incubation period of the FPV and its slow spread with in the house. While vaccinating unaffected birds in cages or pens during an outbreak, vaccination should be started from opposite end of the house to the end from where the outbreak has began. Vaccinate the flocks in adjacent houses also. Mild vaccine strains should be used in laying birds during mid lay period (2nd booster dose) *i.e.* at 42nd weeks of age, if needed.

Mild tissue reactions at the wing web inoculation site called as "Takes" occur in vaccinated birds after 7–10 days of vaccination. These swellings serve as good indicators of successfull immunization to pox vaccine. Vaccination of all birds in a flock should be practiced at same day. Occasionally, pox deases outbreaks in chickens are caused by pox viruses other than fowl pox *viz.* turkey pox, pigeon pox or quail pox. The degree of cross-protection of fowl pox vaccines against other pox viruses varies. Poat-vaccination "Takes" donot occur in previously immunized birds.

Broilers Vaccination

The objective of this vaccination is to prevent the avian pox disease in broilers raised in areas with high pox virus challenge or in areas heavily infested with blood sucking insects including mosquitoes, lice and fleas.

Broiler vaccination with highly attenuated vaccine strains is commonly given in the hatchery by injecting fowl pox vaccines with Marek's disease vaccine. One vaccination is sufficient for protection of broilers life long. However, simultaneous vaccination of pox and MD vaccines, produce interference of development of immunity against MD. The tailor made pox vaccines for administration in day old broiler chicks are only used for the purpose.

REFERENCES

Fallavena, L. C., Canal, C. W., Salle, C. T., Mraes, H. L., Rocha, S. L. and Pereira da Silva, A. B. (2002). Presence of avipoxvirus DNA in avian dermal squamous cell carcinoma. *Avian Pathol.,* 31: 241 – 246.

George, A. W. (2003). *Microbiology Laboratory,* Glencoe press, USA: 327.

Ghildyal, N., Schnitzlein, W. M. and Tripathy, D. N. (1989). Genetic and antigenic differences between fowl pox and quail pox viruses. *Arch. Virol.,* 106: 85 – 92.

Jarmin, S., Ruth, M., Rechard, *E.g.*, Stephen, M. L. and Michael, A. S. (2006). Avian pox virus Phylogenetics: Identification of a PCR length polymorphism that discriminates between the two major clades. *J. Gen. Virol.*, Central Veterinary Laboratories, Weybridge, U. K., 87: 2191 – 2201.

Jordan, F. P., Alexander, M. and Faraghe, D. T. (1996). *Poultry Science* 5th ed. Elsevier, China: 356 – 358.

Mandal, Y. and Johri, P. (2004). *Nutrition and Disease Management of Poultry*. 1st ed. International Book Distributing Co. India: 276 – 278.

Pattison, M. B., McMullin and Alexander, D. (2008). *Poultry Diseases*. 6th ed. Elsevier, India,: 333 – 339.

Rocke, T., Converse, K., Meteyer, C. and Mclean, B. (2005). The impacts of disease in the American White Pelican in North America. *Water Birds*, 28: 87 – 94.

Riper, V. C. III and Forrester, D. (2006). *Avian Pox: Infectious Diseases of Wild Birds*, 06: 131 – 176.

Samour, J. (2004). *Avian Medicine*. 3rd ed. Elsevier, China: 266 – 269.

Schnitzlein, W. M., Ghildyal, N, and Tripathy, D. N. (1988). Genomic and antigenic characterisation of avipoxviruses. *Virus Res.*, 10: 65 – 76.

Smits, J. E., Tella, J. L., Cerrete, M., Serrano, D. and Lopez, G. (2003). An epizootics of avian pox in endemic short-toed Larks (Calndrellarufescens) and Berthelot's pipits Anthusberthelotti)in the Canary Islands. *J. Vet. Pathol.* (Spain), 42: 1 – 59.

Tripathy, D. N. (1986). *Avian Pox*. American Association of Avian Pathologists, 16: 1 – 6.

Tripathy, D. N. (1993). Avipoxviruses. In: *Virus Infections of Vertebrates, Vol 4: Virus Infections of Birds*, McFerran, J.B. and

Tripathy, D. N. and Hanson, L. E. (1976). A smear technique for staining elementary bodies of fowl pox. *Avian Dis.*, 20: 609 – 610.

Tripathy, D. N., Hanson, L. E. and Killinger, A. H. (1973). Immunoperoxidase technique for detection of fowl pox antigen. *Avian Dis.*, 17: 274 – 278.

Tripathy, D. N. and Reed, W. M. (1998). Pox. In: A laboratory manual for the isolstion and identification of Avian Pathogens, 4th ed., Swayne, D. E., Glisson, J. R., Jackwood, M. W., Pearson, J. E. and Reed, W. M. eds. American Association of *Avian Patho*logists, University of Pennsylvania, New Bolton Centre, Kenett Square, PA 19348 – 1692, USA, 137 – 140.

Vegad, J. I., (2008). *Poultry Diseases: A Guide for Farmers and Poultry Professionals*. 2nd ed. International Book Distributing Co. India: 38 – 42.

Winter Field, R. W. and Reed, W. (1985). Avian Pox. Infection and Immunity with Quail, Psittacine, Fowl and Pigeon pox viruses, *Poultry Science*: 65 – 70.

Chapter 8

Egg Drop Syndrome'76

Egg drop syndrome'76 (EDS'76) is an infectious viral disease of egg laying poultry characterized by production of pale, soft- shelled or shell-less eggs by apparently healthy laying birds. The disease is caused by avian Adenoviruses.

Etiology

The avian Adenovirus virus is a non-enveloped, icosahedral particle of 70–90 nm diameter. The particle has 252 capsomeres arranged in 12 triangular faces with 6 capsomeres along each edge. The nucleic acid is linear double stranded dsDNA. Adenovirus replicates in the nucleus resulting in the production of basophilic intranuclear inclusion bodies.

All the adenoviruses are resistant to lipid solvents, sodium deoxycholate, trypsin, 2 per cent phenol and 50 per cent alcohol. They are also resistant to exposure at pH 3 -9 but are inactivated by 1:1000 formalin. The avian adenovirus is more resistant to thermal inactivation. Some strains survive 70°C temperature for 30 minutes, however an F1 isolate is reported to survive for 18 hours at 56°C temperature.

Avian Adenoviruses are devided in to 3 groups *i.e.* I, II and III. Group I or conventional adenoviruses share a common group antigen. These viruses grow readily in avian cell cultures and have been isolated from chickens, turkeys, geese, ducks, quails, pegions, ostriches and other avian species.

Group II adenoviruses include viruses of turkey hemorrhagic enteritis (T.H.E.), Marburg spleen disease (M.S.D.) and group II spleenomegali of chickens. These viruses share a common antigen distinct from group I avian adenoviruses.

Group III viruses *i.e.* egg drop syndrome'76 (EDS'76) disease viruses are widely disributed in waterfowl but can easily infect chickens resulting in the diease EDS'76. The virus comes under the family Adenoviridae in which 2 genera are present. One is Mastadenvirus (mammalian genus) and other is Aviadenovirus genus. A third genus has been recently proposed as Atadenovirus (Benko and Harrach, 1998). EDS'76 virus comes under this new genus. This genus *i.e.* EDS'76 virus partially shares an antigen with F1 adenoviruses (McFerran *et al.*, 1978). EDS virus grows to high titers in duck kidney, duck embryo liver, duck embryo fibroblast cultures and chick embryo liver cells. The virus grows less well in chick kidney cells and grows partially in chick embryo fibroblast cultures. Growth in turkey cell is poor and no growth could be detected in mammalian cells.

Epidemiology and Pathogenesis

After initial entry through the nasal or gastro-intestinal mucosa, local viral replication is followed by a transient viraemia. The principal sites of viral replication are the pouch cell glands and replication occurs to a lesser degree elsewhere in the reproductive tract (Smyth *et al.*, 1988; Yamaguchi *et al.*, 1981). If the embryo is infected or the chick is infected before sexual maturity, the virus remains latent untill the birds become sexually mature. This ensures transmission of the virus to the next generation as virus remains present in and on the eggs for up to 3 weeks (Smyth and Adair, 1988). Viruses are excreted through the cloaca after sheding from the oviducts. Unlike other adenoviruses, EDS'76 virus doesn't excreted from gastro-intestinal tract as the virus has minimal replication in this organ.

All ages of chickens are susceptible although differences in the response may occur. Waterfowls are friquently infected with EDS'76 virus. Quails are susceptible and develop classical clinical signs (Das and Pradhan, 1992).

Three syndromes are associated with EDS. The classical form is seen when primary breeding stocks become infected. Chicks derived from these flocks remain healthy and don't produce antibody until reaching sexual maturity. At some time between the onset of egg laying and peak egg production, abnormal eggs may be produced and the birds produce antibodies. The virus is transmitted vertically through the eggs. Second category is endemic EDS, where the virus subsiquently may infect commercial egg producing flocks and become endemic in that area. This is primarily due to the presence of virus on the exterior of eggs, leading to contamination of tray and trolleys. Infection can also be transmitted horizontly from flocks to flocks by human such as staff and workers dealing with equipements. The third category is the sporadic outbreak which occurs when chickens come in to contact with domestic or wild waterfowls. The contact may be direct or through drinking water.

Signs and Symptoms

The first sign is loss of shell colour in pigmented eggs. This is quickly followed by the appearance of thin shelled, soft shelled or shell-less eggs. The thin shelled eggs often have a rough sand paper like appearance or a grannular roughness at one end. There may be watery albumin. If the obviously affected eggs are removed, fertility and hatchability are not affected.

If the disease develops as a result of reactivation of latent virus, production of usable eggs is reduced approximatly by 40 per cent, this usually occurs when the flock comes in to lay and production remains between 50 per cent to peak. In horizontal/lateral spread of virus, poor egg production may be reported rather than a marked decline. Affected birds appear healthy, but sometimes inappetance, dullness and transient diarrhoea may be seen.

Diagnosis

Selection of the correct specimen is of much importance. No antibody will be produced until sexual maturity in vertically infected chickens. So, proper recommended time to take sample for antibodies is not earlier than 32 weeks of age of birds. Birds are easily marked in cage system than in litter system. The pouch shell gland is the organ of choice for histology, immuno-chemistry or virus isolation. Serum should be collected from the caged birds showing damaged (shell-less/soft shelled) eggs production. Cloacal swabs may also be of great value in virus isolation and identification.

1. Virus Isolation

A 10 per cent suspension is made of the material taken from pouch shell glands and the supernatant is inoculated on to cell culture or embryonated duck eggs. Suitable cells are duck cells, chick embryo liver or chick kidney cells in preferential order. At least 14 days incubation is required after inoculation. If the cells degenerate, the supernatant should be checked for the presence of heamagglutinins using a 0.8 per cent fowl erythrocyte suspension. If agglutination occurs, the isolate can be confirmed by an HI test using specific antiserum.

2. Antigen Detection

Antigen can be detected in the pouch shell glands during the time when defective eggs are being produced using immuno-fluorescent techniques on frozen sections or the avidin–biotin peroxidase technique on formalin fixed tissue sections. In situ hybridization may also be used. Viral antigen can also be detected by PCR and antigen capture ELISA.

3. Serological Tests

The HI (heamagglutination inhibition) test is the method of choice. A 1/10 serum dilution is mixed with an equal volume of a solution containing "4" haemagglutinating units of antigen. The mixture is allowed to react for 15 minutes at room temperature and then one volume of 0.8 per cent fowl erythrocyte suspension is added. Now check for haemagglutination reaction. Other tests like ELISA, Serum Neutralization test and Double immuno-diffusion tests can also be valuable.

Prevention and Control

Basic breeding stock should be free from infection. Sampling at 35 weeks of age in broiler breeders would be acceptable for eradication programme. An inactivated vaccine is useful in preventing EDS'76.

Immunological Conciderations

Antibodies can be detected 7 days post-horizontal infection and peak after 4-5 weeks later. Birds can excrete EDS'76 virus even in the presence of antibodies. Maternally derived antibodies have a half life of 3-4 days. Active antibody production can not be stimulated in birds with MAbs until 4-5 weeks of age, by which time MAbs are virtually undetectable. If the flock as a whole develops immunity at pre-point of lay, effects on egg production will not be seen. Vaccination is typically done using a killed (inactivated) oil-adjuvented vaccine.

Vaccines and Vaccination

The objective of vaccination is to prevent losses in egg production and egg shell quality which occurs with EDS outbreaks in commercial layer and breeder flocks.

Programme

The inactivated (killed) vaccine is given intra-muscularly at least 4 weeks prior to the onset of egg production in layer and breeder flocks. One vaccination is sufficient for stimulating flock immunity life long against EDS. The residue and tissue reaction from oil-adjuvented inactivated vaccines may be detectable many weeks after vaccination. EDS vaccine is only used in that areas where the disease has been diagnosed.

Broilers are not affected by EDS and that's why vaccination is not recommended for broilers.

REFERENCES

Benko, M. and Harrach, B. (1998). A proposal for a new (third) genus within the family Adenoviridae. *Arch. Virol.,* **143**: 829-837.

Das, B. B. and Pradhan, H. K. (1992). Outbreaks of egg drop syndrome due to EDS-76 virus in quail (Coturnix coturnix japonica). *Vet. Rec.,* **131 (12)**: 264-265.

McFerran, J. B., Connor, T. J. and Adair, B. M. (1978). Studies on the antigenic relationship between an isolate (127) from the egg drop syndrome 1976 and a fowl adenovirus. *Avian Pathol,* **7**: 629-636.

Smyth, J. A. and Adair, B. M. (1988). Lateral transmission of egg drop syndrome 76 vims by the egg. *Avian Pathol.,* **17**: 193-200.

Smyth, J. A., Platten, M. A. and McFerran, J. B. (1988). A study of the pathogenesis of egg drop syndrome in laying hens. *Avian Pathol.,* **17**: 653-666.

Yamaguchi, S., Imada, T., Kawamura, T., Taniguchi, T. and Kawakami, M. (1981). - Pathogenicity and distribution of egg drop syndrome 1976 vims (JPA-1) in inoculated laying hens. *Avian Dis.,* **25**: 642-649.

Chapter 9

Chicken Infectious Anemia

Chicken infectious anemia (CIA) is an infectious and contagious viral disease of chickens characterized by aplastic anemia, generalized lymphoid depletion, subcutaneous and intramuscular hemorrhages and immuno-suppression. Immuno-suppression leads to increased mortality because of secondary bacterial complications often observed. The causative agent of CIA is firstly identified in 1979 in Japan and named chicken infectious anemia virus (CIAV). CIA is prevalent in all major chicken producing countries of the world. Chicken is the only natural host for CIAV affecting broilers, breeders and layers. The disease is transmitted in both ways horizontally as well as vertically. But vertical transmission appears to be the most important means of dissemination. Vertical transmission occurs following the infection of hens in lay.

CIAV is a spherical or hexagonal, non-enveloped virus with a diameter of 23.5–25 nm having icosahedral symmetry, single stranded (ss) DNA containing genome belonging to the genus Gyrovirus of the family Circoviridae (Pringle, 1999). No significant antigenic differences have been recognized among various CIAV strains using polyclonal antibodies. There is only one serotype of CIAV worldwide (Adair, 2000).

The structure of CIAV confers remarkable chemical and thermal stability. Virus has the ability to resist inactivation by various physico-chemical agents *viz.* exposure to pH-3, lipid solvents (ether and chloroform), acetone and treatment for 2 hours at 37°C temperature with 5 per cent solutions of many commercial disinfectants. Treatment with 1 per cent gluteraldehyde for 10 minutes at room temperature, 0.4 per cent of beta-propriolactone (BPL) for 24 hours at 4°C temperature and 5 per cent formaldehyde for 24 hours at room temperature have been recommended for complete inactivation of the virus (Yuasa, 1992). Sodium hypochlorite in 1 per cent concentration can also inactivate the virus. The virus has been shown to withstand temperature of 70°C for 1 hour and 80°C for 15 minutes. Heating at 100°C for 15 minutes completely inactivates the virus.

Epidemiology

CIAV has worldwide distribution in major poultry producing countries including India (Kataria *et al.*, 1999). Chicken is the only natural host for the virus but quail may be affected. Vertical as well as horizontal modes of transmission are involved in the spread of CIAV among chickens resulting in clinical and sub-clinical infections, respectively.

Vertical transmission occurs when breeders and layers are without antibodies to CIAV or with no exposure to the virus or become infected as they come in to lay. Clinical disease doesn't occur in breeders and there is no apparent adverse effect on egg production, hatchability or fertility but the virus is passed vertically to the progeny chicks which in turn develop clinical disease.

Horizontally acquired infection, usually occurs in chickens lacking MAbs to CIAV. Infection usually contracted by direct or indirect contact from virus surviving in poultry houses between flocks, from virus excreted by a small number of vertically infected hatch mates or through ingestion of materials contaminated with infected feces.

Vertical transmission being the major cause of clinical disease has more importance than that of horizontal transmission. Sub-clinical infections in chicks above 3 weeks of age are much more common. Outbreaks of CIAV are sporadic in the field (Todd, 2000).

Mortality peaks within the 3rd week of life of chickens and ranges between 5–10 per cent, may up to 60 per cent and morbidity varies between 20– 60 per cent. In heavily infected flocks, there may be a second peak of mortality in 30–34 days old birds possibly due to contracting horizontal infections.

Viral Replication

After adsorption and penetration, the virus probably enters the target cells and multiplies in the nucleus by circular dsDNA replicative intermediate form (rolling circle model). In chickens, CIAV appears to replicate primarily in hemopoietic precursor cells in the bone marrow and thymic precursors (lymphocytes) in the thymus cortex, where it leads to cytolytic infection resulting in to cell death by apoptosis (Jeurissen, *et al.*, 1992).

Three putative viral proteins *viz.* VP1, VP2 and VP3 have been well characterized. VP1 AND VP2 are involved in antigenicity of the virus and formation of neutralizing antibodies in CIAV infected cells. VP3 (apoptin) is the major protein found in the infected cells which induces apoptosis in specific lymphoid cells, the chicken thymocytes and chicken lymphoblastoid cell lines (Jeurissen, *et al.*, 1992).

Pathogenesis

The primary target cells appear to be hemopoietic precursor and thymopoietic precursor cells in the bone marrow and thymus cortex, respectively in early cytolytic infection phase at 6–8 DPI (days post-infection). CIAV antigens have been detected in cells of bone marrow, thymus and spleen at 3–4 DPI and subsequently in other tissues *viz.* liver, proventriculus, duodenum, lungs, kidneys and heart indicating its wide distribution throughout the body. No antigen has been detected after 26 DPI (Adair *et al.*, 1993b). CIAV replicates in lymphocytes causing destruction of thymic lymphocytes and is directly cytotoxic for bone marrow hemopoietic precursors leading to immuno-suppression and transient severe anemia, respectively (Adair, 2000).

The infection of thymocytes with CIAV causes chromatin aggrigation, fragmentation of cellular DNA in to oligo-nucleosomes, karyorrhexis and cell death by apoptosis. Nuclear inclusions are exclusively VP3 induced apoptotic bodies (Jeureissen, *et al.*, 1992). Eosinophilic intranuclear inclusion bodies have been found in altered cells especially in bone marrow and thymus. Hemorrhages associated with CIAV infection are chiefly due to destruction of thrombocytes leading to impaired clotting mechanism of blood. Depletion of grannulopoietic tissues is responsible for the lack of mounting inflammatory responses in the immuno-suppressed chicks. So, pathogenesis of CIAV infection is strongly due to hampered humoral immunity.

Gross Lesions

Transient severe bone marrow aplasia and pancytopaenia with reduction in hematocretic values are primarily seen. There is normochromic anemia. Blood becomes watery, plasma becomes paler and blood clotting time is increased. Bone marrow characteristically changes from red to pale-yellow/white color with fatty consistency. Femur is most commonly evaluated for this change. Liver becomes pale, discolored and enlarged. Lymphoid organs like Bursa of Fabricius, thymus and spleen are atrophied. Focal necrosis of liver, kidneys, spleen, erosion of gizzard, skin necrosis on wings and consolidation of lungs are also marked. Hemorrhages in the mucosa of proventriculus, subcutaneous and muscular hemorrhages within the wing tips are associated with severe anemia (Yuasa and Imami, 1986).

Microscopic Lesions

Mild to severe depletion of erythroid and myloid cells are seen. Bone marrow gives a watery texture and characteristic yellow color. Atrophy of lymphoid follicles is seen in Bursa of Fabricus with hydropic epithelial degeneration. In the liver, kidneys, proventriculus, duodenum and ceacal tonsils, the lymphoid foci are depleted of cells, making them smaller and lesser dense than those in unaffected birds. Liver cells are swollen, hepatic sinusoids are dilated and the lymphoid foci are depleted of cells. Vacuolar degeneration is seen in hepatocytes. Small eosinophilic intranuclear inclusion bodies are seen in altered cells which are pathognomonic.

Signs and Symptoms

Clinical signs appear after an incubation period of 10–14 days and include paleness, anemia, weakness, anorexia and stunted growth. Anemic conditions are grossly seen on combs, wattles, eyelids and legs.

Diagnosis

It is based on clinical signs and gross pathological lesions tentatively. However, confirmatory diagnosis needs isolation and identification of the agent (McNulty, 1998). The virus can be isolated by inoculation of the suspected contaminated/ infected material in the susceptible embryonated chicken eggs, cell culture, one day old SPF (specific pathogen free) eggs and/or susceptible immuno-suppressed chicken. CIAV can be isolated from virtually all tissues of infected chickens. Maximum virus titer can be detected around 7 DPI (days post-infection). Liver is the best source followed by thymus, bone marrow, spleen, bursa, lungs, heart, muscle and rectal contents (McNulty, 1998). Inoculation of the susceptible 5–7 days old embryonated chicken eggs via yolk sac route can be used for isolation of CIAV from all parts of the embryo but not from yolk or CAM (chorio-allantoic membrane). Embryo becomes small, hemorrhagic and edematous. CPEs (cytopathic effects) are not readily observed in few first passages because non-infected cells of cell culture outgrow the infected cells. Inhibition of cell growth, enlarged, swollen and misshapen cells in cell cultures are seen.

Virus can be detected by electron microscope. DNA of virus can be detected by PCR technique. Nucleic acid hybridization, restriction enzyme mapping are other methods to detect DNA. VNT, IP, IFT and ELISA are serological methods to detect antibodies in serum.

Prevention and Control

Control measures are mostly directed at limiting vertical transmission and subsequent clinical disease outbreaks. Prevention by the use of vaccine in parent flocks is aimed towards sero-conversion before coming the birds in lay. Sound management, hygiene and strict biosecurity practices are of immense help in preventing young chicks from early exposure to CIAV and other lymphocytocidal agents like IBDV, MDV and certain fungal toxins.

Vaccination with live vaccine should be done at the age 4 weeks prior to start of egg laying, so that the vaccine virus cannot come in to eggs, thus preventing the vertical transmission of infection to the progeny chicks. Conventional approaches for development of a vaccine present difficulties due to the inability of the virus to grow up to high titers in chicken embryos/cell cultures (*in vitro*) and the unavailability of naturally occurring apathogenic isolates of CIAV (Koch, *et al.*, 1995).

Immunological Considerations

CIA can be effectively prevented by humoral immune response in the immune-competent host or with passively transferred MAbs (Yuasa *et al.*, 1985). There is no much literature about the role of CMI (cell mediated immunity) in CIAV infection. However, its role in protection against CIAV is not ruled out. MAbs are protective in chicks up to first 2–3 weeks of age after which, the age related resistance develops against clinical disease but not against infection (Pages *et al.*, 1997). Because of the passive protection through MAbs in most of the broiler flocks against CIA, only subclinical disease is produced. This subclinical disease is produced mostly at an age of 3–6 weeks after disappearance of MAbs. Such type of infection affects growth and health of birds negatively, although significant effect has not been seen on mortality. Severity of disease is related to the virus dose, age of birds at infection, the MAbs status oh chicks and the route of infection (vertically or horizontally). Bursal atrophy is an important risk factor for the development of CIA.

Because of the pathological effects of CIAV, the virus is believed to be a potent immuno-suppressive agent in young susceptible chicks. When the virus is transmitted by transovarian route from infected hens to progeny chicks, CIAV can cause a severe disease resulting in decreased resistance and enhanced susceptibility to a wide range of viral, bacterial and fungal infections (Rosenberger and Cloud, 1998). Subclinical infections in chickens of age more than 3 weeks can also result in immuno-suppression (Adair, 2000). CIAV puts a destructive effect on both primary and secondary lymphoid tissues and specially suppress the population of both helper (CD4+) and cytotoxic (CD8+) T-lymphocytes in thymus leading to immuno-suppression (Adair, 2000). There is marked damage to hemopoietic and lymphopoietic tissues *viz.* stem cells in bone marrow and precursor T-lymphocytes in thymus. The bursa, spleen and other lymphoid organs are also less severely depleted of lymphoid cells. Inhibition of interleukins (IL-1, IL-2 *etc.*) and interferon (IFN) production adversely effects molecular immuno-regulatory responses on cytotoxic activities of macrophages, cytotoxic T-lymphocytes, natural killer (NK) cells and expression of surface receptors (Adair, 1993).

CIAV infection causes depression of immune response against several vaccine viruses *viz.* NDV, IBDV, MDV, ILTV and FPV leading to vaccination failures. Vaccination reactions and/or aggravation of the residual pathogenicity of attenuated vaccine viruses may lead to emergence of variant viruses.

Vaccines and Vaccination

Purpose of Vaccination

The purpose of vaccination is to prevent vertical transmission of CIAV from breeder hens (infected during egg production) to their progeny chicks. Progeny chicks may develop severe anemic and immuno-suppressive diseases as a result of CIAV infection.

Vaccination

The objective of live attenuated vaccination programme is to design vaccination for use in breeder flocks to produce immunity against CIAV infection prior to the onset of egg production.

Vaccination of breeder flocks against CIAV with live attenuated vaccine as drinking water administration is practiced 4 weeks prior to the onset of egg production *i.e.* at 12th week of age, as in well managed flocks, egg production starts at 16th week of age of birds. The generally used vaccine virus strain is Cux-1 or local strain. Single vaccination with live vaccine using Cux-1 strain or local strain is sufficient for life time in breeder hens. The vaccine rules out the vertical transmission of CIAV in progeny chicks and thus no disease occurrence. Where the live vaccines are not available, the litters from CIAV positive breeders can be used as controlled exposure practices. However, controlled exposure of breeder pullets to litter from CIAV positive breeders can result in the spread of other pathogenic viruses or bacteria. Breeder flocks which are not sero-positive to CIAV by 12 weeks of age should only be immunized. CIAV vaccines should only be used in areas where the diseased has been diagnosed. CIAV negative breeder hens and young birds (less than 6 weeks of age) of the areas from where disease has not been reported should not be exposed to the live vaccine or contaminated litter.

REFERENCES

Adair, B. M. (2000). Immuno-pathogenesis of chicken anaemia virus. *Dev. Comp. Immunol.*, 24: 247 – 255.

Adair, B. M., McConnell, C. D., McNeilly, F., McNulty, M. S. and Caudert, F. (1993). Effect of chicken anaemia virus on macrophage and lymphocyte functions in chickens. *Avian Immunology in Progress*, 219 – 223.

Jeureissen, S. H. M., Wagennar, F., Pol, J. M. A., Van Der Eb A. J. and Noteborn, M. H. M. (1992). Chicken anaemia virus causes apoptosis of thymocytes after in vivo infection and of cell lines after *in vitro* infection. *J. Virol.*, 66: 7383 – 7388.

Kataria, J. M., Suresh, R. P., Verma, K. C., Toroghi, R., Kumar, N. S., Kataria, R. S. and Sah, R. L. (1999). Chicken infectious anaemia (CIA) in India: detection of agent by polymerase chain reaction and transmission study. *Ind. J. Comp. Microbiol. Immunol. Infect. Dis.*, 20: 91 – 95.

Koch, G., Van Roozelaar, D. J., Verschuren, C. A. J., Van Der Eb A. J. and Noteborn, M. H. M. (1995). Immunogenic and protective properties of chicken anaemia virus proteins expressed by baculovirus. *Vaccine*, 13: 763 – 770.

McNulty, M. S. (1998). Chicken anaemia virus. In: *A laboratory Manual for the Isolation and Identification of Avian Pathogens*. 4th Edn. D. E. swayne, J. R. Gilson, M. W. Jackwood, J. E. Pearson, W. M. Pearson, W. M. Reed (Editors), American Association of Avian Pathologists (AAAP), Univ. of Pennsylvania, Philadephia, P. A., pp: 146 – 149.

Pages, M. A., Saubi, N., Artiges, C. and Espuna, E. A. (1997). Experimental evaluation of an inactivated vaccine against chicken anaemia virus. *Avian Pathol.*, 26: 721 – 729.

Pringle, C. R. (1999). Virus Taxonomy at the XIth International Congress of Virology, Sydney, autralia. *Arch. Virol.*, 144: 2065 – 2070.

Rosenberger, K. K. and Cloud, S. S. (1998). Chicken anaemia virus. *Poult. Sci.*, 77: 1190 – 1192.

Todd, D. (2000). Circovirus: Imminosuppressive threats to avian species: A review. *Avian Pathol.*, 29: 373 – 394.

Yuasa, N. (1992). Effect of chemicals on the infectivity of chicken anaemia virus. *Avian Pathol.*, 21: 315 – 319.

Yuasa, N and Imai, K. (1986). Pathogenicity and antigenicity of eleven isolates of chicken anaemia agent CAA. *Avian Pathol.*, 15: 639 – 645.

Yuasa, N., Imai, K. and Tezuka, H. (1985). Survey of antibody against chicken anaemia agent (CAA) by an immunofluorescent antibody technique in breeder flocks in Japan. *Avian Pathol.*, 14: 521 – 530.

Chapter 10

Avian Mycoplasmosis

Mycoplasma gallicepticum (MG) and Mycoplasma synoviae (MS) remain the most important mycolasma infections affecting chickens. First one causes chronic respiratory disease (CRD) and later one causes infectious synovitis. MG and MS produce diseases in chickens, turkeys and some other avian species. MG is the most economically significant mycoplasma pathogen of poultry and has a world-wide distribution. As common in other mycolpasma, MG is also minute in size with minimal genetic information and with a total lack of cell wall (Razin, 1992). The most dramatic disease presentation of MG is chronic respiratory disease in meat type (broiler) birds. Transmission of MG in vivo from infected breeder birds to progeny chicks is the major route of dissemination of the infection and is also of prime consideration for international trade. MS was noticed initially as an agent of synovitis and the disease was called infectious synovitis in broilers. However, MS was later frequently isolated from air sacculitis lesions in broiler flocks which were free from MG infections. *M. meleagridis* is a specific pathogen of turkey.

Etiology

Approximatelly 25 named mycoplasma – the mollicutes (beloging to the genera Mycoplasma, Acholeplasma and Ureaplasma) have been isolated from avian species. The presence of only a minimal amount of genetic information accounts for the complex nutritional requirements of these organisms reflected in an obligate parasitic mode of life, with a high degree of interdependence between the mycoplasma and the host. Mycoplasma organisms tend to be highly host specific.

The primary habitates of the mycoplasma are mucosal membranes of the respiratory tract, uro-genital tract, eyes and joints. Adhesions of mycoplasmas to host cell is a pre-requisite for successful colonization and ensuring pathogenesis (Krause, 1996). Some species may penetrate cells and survive intracellularly (Razin *et al.*, 1998).

Epidemiology

MG infection occurs naturally in chickens and turkeys, the transmission of which occurs by the following two major routes:

1. Vertically (*in vivo*), from an infected breeder flock to the progeny chicks.
2. Horizontally by direct or indirect contact of susceptible birds with infected carriers or contaminated debris.

In addition to domestic poultry, MG has been reported frequently in other gallinaceous birds, either free ranging or maintained in captivity and sporadically in birds of other genera (Lay and Yoder, 1997). Natural infections with MG are also reported in a variety of game birds. Isolation of MG from respiratory tract of ducks with no clinical sign has also been reported (Bencina *et al.*, 1988). MG has also been isolated from embryonated duck eggs from the infected flocks.

Horizontal infection of MG is readily transmitted through contact with infected birds most probably by the airborne route with organisms excreted from the respiratory system of infected birds through aerosol. However, no published evidence is available to demonstrate the role of airborne transmission of MG under field conditions, although circumstantial evidence is often found. MG could be re-isolated from several innert materials 2 days post contamination and also survived on human hairs for 3 days (Christensen *et al.*, 1994). The susceptibility of chickens to MG or MS is significantly increased in the presence of high concentrations of ammonia fumes or dust which leads to reduction in normal physiological activities of the respiratory mucus membrane.

Pathogenesis

At present, very little is known about the pathogenesis of mycoplasma in poultry.

Description of Disease

The most economically significant mycoplasma pathogen of poultry is MG (Ley and Yoder, 1997). The major concideration for international trade is the necessity and the ability to determine the MG status of imported birds, day old chicks and hatching eggs. In most cases, the MG status of the progeny chicks is determined by the breeder flocks from where they are obtained.

Infection with MG may have a wide diversity of clinical manifestations, of which "chronic respiratory disease" and down grading of carcass in meat-type birds are probably the most dramatic (Kleven, 1998). Environmental and stress factors such as temperature, ammonia concentration, age and type of birds are among the factors that influence the mycoplasmosis (Ley and Yoder, 1997). Moreover, infection of MG may cause a marked reduction in feed consumption efficiency with significant economic impact, even in the absence of overt clinical signs. Loss of production in laying birds may also occur as a result of MG infection (Glisson and Kleven, 1984). Early infection of pullets with MG can protect the birds against the pathogenic effects of later infections (Fabricant, 1975). This is the principle underlying the successful use of live MG vaccines in layer flocks as it widely practiced in the multi-age layer operations in U.S.A. and elsewhere (Whithear, 1996). MG has been implicated in salpingitis and other pathologies of the reproductive system, although it is sole or principal cause of decreased egg production is not clear (Nunoya *et al.*, 1997).

MG is transmitted in ovo. Infection of the embryo by MG and transmission to the progeny chicks (vertical transmission) probably occur as a sequel to acute respiratory infection due to contiguity of the abdomenal air sacs with the oviducts (Roberts and McDaniel, 1967). The highest frequency of transmission is found during the acute phase of the disease when MG concentration in the respiratory tract reaches a peak (Lin and Kleven, 1982). The presence of MAbs to MG in the embryonated eggs have been found to reduce the in vivo pathogenecity of infection, increasing the probability of survival of the infected embryoes (Levisohn *et al.*, 1985). However, MG infection may weaken the embryo, resulting in difficult hatching (piped embryoes) or low quality chicks. The trachea serves as reservoir organ for MG infection. Concentration of MG in trachea is highest at the acute stage of infection. However, the tracheal infection persists in the presence of humoral or local antibodies (Yagihashi and Tajima, 1986).

Signs and Symptoms

MG causes respiratory disease, reduced feed conversion and decreased egg production efficiency. So, it is one of the costliest diseases confronting the poultry industry (Yoder, 1991). In nut shell, clinical signs associated with MG infections include watery eyes, sneezing, coughing rales, unilateral or bilateral swollen infraorbital sinuses, ataxia, lameness, torticolis, arthritis, synovitis and mild to sever conjunctivitis. Reduced body weight gain, poor feed conversion ratio (F.C.R.), increased carcass condemnation, reduced egg production and decreased hatchability of eggs are other signs.

MS causes respiratory disease and synovitis in growing birds, although subclinical infections are common (Kleven *et al.*, 1991). MS is common in multiple-age commercial layer flocks (Kleven *et al.*, 1991) but it is reported to have no significant impact on egg production (Mohammed *et al.*, 1987). Clinical signs associated with MS infections are swollen joints, lameness and respiratory signs particularly in breeders. However, it will be premature to conclude just based on clinical signs that the flocks are suffering from mycoplasma infections.

Lesions

Lesions that have been reported in single or mixed infections of MG are mucous and/or catarrhal exudates in the nostrils, sinuses, trachea, bronchi, lungs and air sacs. Mild to sever air sacculitis can also be seen. Additionally pneumonia, fibrinous perihepatitis and fibrinous pericarditis have also been reported in association with secondary bacterial infections. Salpingitis may also be found.

MS infection associated lesions may be synovitis, teno-synovitis, air sacculitis and bursitis.

Diagnosis

The basis of control programmes to be adopted lies on serological diagnostic methods, both screening and confirmatory with reactors. Final confirmation is done by isolation and identification of the organism. Diagnosis is made on flock basis. The presence of one or more infected bird(s) in the flock sample constitute(s) an infected flock.

Diagnosis of MG infections in poultry breeder flocks is often performed in the absence of overt clinical signs. Screening for infection is usually done by Rapid Serum Plate Agglutination Test (RSPAT) with commercially available stained antigens. RSPAT is highly efficient in detecting immunoglobulins of Ig M class, which are the earliest response to mycolpasma infection (Kleven, 1975). Although, the test is rapid, highly sensitive and relatively inexpensive, care must be taken to perform the test according to the instructions of the antigen manufacturer using appropriate sera (Kleven, 1996). The greatest disadvantage of RSPAT is low specificity, with false positive reactions and frequent cross-reactions encountered. Serological detection of MG may be complicated by co-infections of flocks with MS, due to serological cross-reactions between the two mycoplasm species (Avakian and Kleven, 1990). In addition, non-specific serological reactions are commonly detected after use of inactivated vaccines (Glisson *et al.*, 1984).

Traditionally, the test of choice for confirmatory serology is haemagglutiation-agglutination (HI) test, which can be performed with fresh culture of a haemagglutinating test strain of MG (Kleven *et al.*, 1996) or with standardized preserved antigen (APHIS, 1997). Diagnostically significant titres in the HI test may not be detected until 3 or more weeks after infection. However, the test is highly specific, even to the level of differentiation among strains (Kleven *et al.*, 1988). Commercial ELISA kits are widely available and are increasingly used for serological confirmation (Kempf *et al.*, 1994).

Ig G passes from maternal circulation in to the egg yolk. Transferred Ig M has also been detected at low levels in the yolk but at more significant levels in egg white (Yamamoto *et al.*, 1975). However, only maternally derived Ig G, transferred through egg yolk remains present in chick's circulation (Rose and Orlans, 1981). This Ig G passes from yolk to the embryo during the last 5 or 6 days of embryonic development. In the chicks for approximately after 2 days post hatching the antibody (Ig G) titre declines. The ELISA is usually recommended for testing of yolk samples in fertile or infertile eggs and testing MAbs in the chick's serum (Brown *et al.*, 1991).

In principle, the presence of MG organisms can be confirmed by isolation in mycoplasma media or by detection of the specific DNA by PCR technique. Isolation is still considered as the "gold standard"method but the existance of circumstances where MG may be present but cann't be isolated even by the most reliable techniques, is now fairly well accepted. Details of methods for culture and identification of MG may be found in the O.I.E. mannual (Kleven *et al.*, 1996).

PCR represents a rapid and sensitive alternative method over traditional cultural methods which requires specialized media and reagents. Although, PCR method is time consuming (Kempf, 1998) but now has become one the test of choice.

However, detection of DNA from non-viable organisms, for instance after antibiotic treatment, is a possible drawback to the PCR method (Kempf *et al.*, 1994). It is further important that in case of mixed infection by several mycoplasma species (including certain saprophytic mycoplasma), the PCR result may be false positive for MG.

Due to the high sensitivity of PCR test, care should must be taken to avoid false positive reactions due to extraneous DNA but it can be prevented or detected by appropriate controls. A critical control in the use of PCR for diagnosis is the inclusion of an internal control to avoid false negative results because of the presence of inhibitory substances in the reaction mixture. Amplification of the internal control amplicon, which can be readily differentiated from the target DNA amplicon, indicates that there are no inhibitors of the PCR reactions.

A recent innovation in diagnosis is the development of molecular typing method for differentiation of MG strains. The most commonly used molecular typing method is random amplification of polymorphic DNA (RAPD), a PCR based technique which gives a unique strain fingureprint (Fan *et al.*, 1995). It can be able to differentiate even MG vaccine strains and field strains present in natural infections (Charlton *et al.*, 1999). Current research is attempting to develop rapid molecular methods for specific detection of MG biotypes, such as the live vaccine strain, as has been successful for "F" vaccine strain (Khan *et al.*, 1989). The new refinement of PCR, *i.e.* RAPD is being used to distinguish among "ts-11" and "6/85" strains (Ley *et al.*, 1997) and "F-strain, ts-11 and 6/85 strains (Fan *et al.*, 1995).

Sampling, sample transport and sample processing for MG testing are highly critical stages in diagnosis. Sampling for MG in live birds is usually done from the trachea or the choanal cleft. During the acute phase of infection, 20–30 samples from individual birds for isolation of mycoplasma are usually sufficient, where as a larger sample size may be necessary during the chronic stage of infection. Prewetted swabbing technique is successful from trachea and choanal clefts. Isolation has been also successfull from brain or eyes of the fowl with relevent clinical signs (Nunoya *et al.*, 1995) in addition to bile of the infected birds (Bencina *et al.*, 1991).

Several rapid methods for extraction of mycoplasma DNA from infected cells have been adopted for sample preparation to be used in PCR (Lauerman, 1998). Samples for PCR are often pooled (3-5 tracheal swabs per PCR reaction) to increase sample size and reduce the cost of testing. However, pooling of samples may increase the possibility of reaction inhibition by substances that may be present in the mucous or other tissue fluids, thus decreasing the sensitivity of the PCR test.

When pooling of large number of samples has to be done, purification of the DNA may be necessary using standard methods or rapid commercial kits. Testing for MG in the embryonated eggs or progeny chicks by culture method is not recommended as a routine method for determining the status of the flock.

When using RSPAT, non-specific reactions may occur in some flocks which are inoculated with an inactivated tissue cultured IBDV (infectous bursal disease virus) vaccine. This is caused by cross-reactions between the antibody for serum component in the tissue culture of the IBDV vaccine and the serum factor adsorbed on the mycoplasma antigen. It has been recommended that, to avoid these cross-reactions, a drop of fresh horse or pig serum should be added to the test serum in RSPAT (Hidaka *et al.*, 1987).

Diagnostic Standards in Various Serological Tests for MG and MS infections of Chicken

Sl.No.	Serological Tests	MG			MS		
		Positive	Suspicious	Negative	Positive	Suspicious	Negative
1.	*WBPAT	< 1 min	1~2 min	> 2 min	< 2 min	-	> 2 min
2.	*RSPAT	< 1 min	1~2 min	> 2 min	< 1 min	1~2 min	> 2 min
3.	*TA	>/= 1:20	1:10	< 1:10	>/= 1:20	1:10	< 1:10
4.	*HI	>/= 1:10	1:5	< 1:5	>/= 1:10	1:5	< 1:5

WBPAT = Whole blood plate agglutination test; RSPAT = Rapid serum plate agglutination test; TA = Tube Agglutination; HI = Heamagglutination-inhibition.

Source: Ando *et al.*, 1965; Kuniyasu and Ando, 1966; Sato, 1976.

Prevention and Control

In most of the countries, control programmes for MG and MS are based on maintenance of freedom from infection in commercial breeding stock. Both MG and MS are egg transmitted pathogens. So, by breaking the cycle of vertical transmission, it is possible to eliminate these infections from breeder stock. The basic concept of MG/MS control depends upon the following measures:

1. A high level of biosecurity of breeder flocks with production in single-aged, all-in/all-out farms.

2. Routine monitoring by serological testing backed up by rapid and specific confirmatory tests.

3. Immediate slaughter of infected breeding flocks to prevent transmission of mycoplasma in the progeny chicks.

In general, purchasers of fertile hatching eggs or day old chicks/poults require certification that the source breeding flock is free from MG infection. Veterinary authorities of importing countries should require the presentation of an international veterinary certificate attesting that the birds:

1. Showed no clinical sign of avian mycoplasmosis on the day of shipment.

2. Come from an establishment free from avian mycoplasmosis.

3. Were kept in a quarantine station for 28 days prior to shipment and were subjected to a diagnostic test for avian mycoplasmosis with negative results on two occasions, first at the beginning of quarantine and second at the end of quarantine *i.e.* at 28th day.

The above international veterinary certificate for day old chicks and for hatching eggs of chicken/turkeys should must comply with the standards reffered in terrestrial animal health code, chapter 6.4 of OIE (at web.oie.int/eng/normes/mcode/en_chapitre_1.6.4.htm).

MG and MS are known to be susceptible to several antibiotics including macrolides, tetracyclines, fluoroquinolones and others (Bradbury, et al., 1994). But as mycoplasma is devoid of cell wall, penicillins, cephalosporins and other antibiotics which act by inhibiting cell wall biosynthesis are ineffective. MG may develop resistance against commonly used antibiotics (Zanella et al., 1998). However, medication should not be relied upon to eliminate MG and MS infection from an infected flock and should be regarded as a method for short-term amelioration of signs and economic effects, rather than as a long term solution to the problem.

In case, where control of MG/MS infections are not feasible, vaccination with either inactivated or live vaccines may be an option. Vaccination is most commonly used in commercial layer pullets to be placed on multi-age commercial egg production sites and in some instances, in broiler breeder pullets.

Procedures for Establishment of Mycoplasma Free Breeding Flocks

Procedures have been formulated for the establishment of MG free breeding flocks. For the selection of a breeding flock, from which the hatching eggs will be used to establish mycoplasma-free stock, two serological tests are performed at one month intervals. If the antibody titres of this flock in both the tests remain the same, the flock is selected and antibiotics effective for MG (e.g. tylosin) are administered. Such treatments prevent or reduce the production of MG contaminated eggs for a period of approximately one month. Furthermore, if necessary, the hatching eggs are treated with antibiotics to eliminate transmission of mycoplasma, using the egg dipping method or injection in to the eggs. Chicks resulting from the treated eggs are devided in to small flocks (fewer than 500 chicks each) and are raised in strict isolation with regular serological monitoring. Flocks including seropositive chickens are discarded. These procedures should be repeated over two generations. When complete elimination of the infection has been achieved, the flocks can be determined as MG free breeding stocks (Minamimoto et al., 1968).

For the establishment of MS free breeding flocks, the egg heating procedure seemed more promising than the egg dipping method (Yoder, 1970). Eggs at room temperature (24.4°C) should be heated in an incubator over 13.5 hours to reach a temperature of 46.6°C to inactivate MS contamination. The temperature should be gradually raised to the critical point using a special thermal regulator developed by Murayama (Murayama, 1984). In this method, the hatchability of heated eggs is reduced by 4-15 per cent (Murayama et al., 1977).

Immunological Considerations

Immune mechanisms operating in MG/MS infections are not fully understood. It is possible that different mechanisms may operate for different routes of challenge and for resistance to infection compared with recovery from disease. The importance of immunoglobulin (Ig) has been demonstrated because bursectomised chickens have lower resistance to MG than normal chickens or thymectomised chickens (Lam and Lin, 1984). However, a lack of correlation between levels of specific circulating antibodies and protection has been well documented (Talkington and

Kleven, 1985; Whithear *et al.*, 1990). It is likely that antibodies in the respiratory tract are important with the predominating Ig class in respiratory secretions being IgG (Yagihashi and Tajima, 1986; Avakian and Ley, 1993) instead of Ig A. A protective role for cell-mediated immunity (CMI) has been discussed but the evidence for the significance of CMI is largely circumstantial (Adler, 1976).

As with MG immunity, the MS immunity is also bursal dependent (Kume *et al.*, 1977; Vardaman *et al.*, 1973), although protection is not correlated with level of serum antibdies. Antibodies in respiratory secretions may be more important.

In general, chickens infected with MG or MS produce no clinical symptoms. However, it is known that chickens infected subclinically with MG or MS can develop CRD or severe respiratory lesions in complication with other infectious agents. When chickens were inoculated only with MG, no clinical signs were observed. However,in chikens infected simultaneously with MG and N.D. virus (B1 strain), severe respiratory symptoms were observed (Nonomura, *et al.*, 1971).

Vaccines and Vaccination

There are two types of vaccines used to prevent mycoplasmosis in poultry.

1. Killed/Inactivated bacterins
2. Live vaccines

1. Killed/Inactivated Bacterins

MG bacterins are used commercially in several countries which contain inactivated organisms suspended in either aqueous emulsion (Hildebrand *et al.*, 1983) or Aluminium hydroxyde adjuvants (Yagihashi and Tajima, 1986). Bacterins have the advantage that they are non-infectious and thus pose no risk of cross-infection to other stock or of reversion to virulence which are potential problems with live vaccines. However, these bacterin vaccines are expensive, require large amount of antigens and necessisate individual bird handling during vaccination as intramuscular or subcutaneous routes. The site of injection is important. Chickens vaccinated subcutaneously at the base of the skull, develop a transient edema around the eyes, whereas those vaccinated subcutaneously midway or lower in the nape of the neck showed no obvious adverse reaction (Hildebrand *et al.*, 1983). Intramuscular injection of vaccine in breast muscles results in losses during processing. So, the intramuscular injection in upper leg is an alternative. However, granulomatous cellulitis may develop (Droual *et al.*, 1993).

Chickens vaccinated with oil-emulsion bacterins (with a laboratory prepared vaccine) at one day of age demonstrated little protection, those vaccinated at 7 days of age, varriable protection and those vaccinated at 11 days or older age, showed significant protection (Yoder *et al.*, 1984; Glisson *et al.*, 1984). The cause of it may be the presence of MAbs in early life of chicks leading to neutralization of administered antigens. Broiler breeders vaccinated at 8th and 12th weeks of age showed only some protection following natural challenge (Gallazzi *et al.*, 1985).

2. Live Vaccines

Presently, at least three live MG vaccines are recommended for commercial use in various parts of the world. These include the "F strain" which is a naturally occuring strain of moderate virulence to chickens and high virulence to turkeys. Strains 6/85 and ts-11, which are artificially attenuated strains of low virulence. Strain MS-H is an attenuated strain of MS, which is currently undergoing evaluation.

A live mycoplasma vacine strain must multiply in the bird to an extent sufficiently to stimulate long-term protective immunity but unable to cause disease or spread to other susceptible birds.

Methods of Administration

"F strain" is usually administered by aerosol spray, but has also been given in drinking water or by intranasal instillation (Levisohn and Kleven, 1981). This strain was found to be sufficiently stable in a variety of diluents to allow the strain to be used in drinking water (Kleven, 1985). It is not suitable to be used in turkeys because of its too virulence to turkeys (Lin and Kleven, 1982) while ts-11 is apathogenic for turkeys (Whithearet al., 1990). The ts-11 strain was developed by chemical mutagenesis of an MG isolate in Australlia and selected as temperature sensitive (ts) mutant and is avirulent to chicken and turkeys. It induces long lived immunity (Whithear, 1996).

The recommended route of administration for "strain 6/85" is by aerosol spray (Evans and Hafez, 1992), whereas "strain ts-11" is administered by eye drop (Whithear *et al.*, 1990). Dosage is important with these two live attenuated vaccine strains. A dose of 10^7 organisms of a fresh culture provided significant protectionin case of "strain 6/85" (Evans and Hafez, 1992). The recommended eye drop dose of "ts-11 strain" is more than/equal to (>=) $10^{7.7}$ organisms (K. G. Whithear, unpublished findings).

"Strain 6/85" is supplied as a lyophilized product, whereas "ts-11 strain" is currently distributed as a culture frozen at $-70°C$ temperature. The "ts-11 strain" vaccine is being used in the field in flocks of 2 weeks to 16 weeks of age.

"Strain MS-H" is administered and distributed in the same way as "ts-11 strain" of MG *i.e.* by eye drop. In Australlia, "MS-H strain" vaccine is commonly administered in to the alternate eye at the same time as vaccination with "strain ts-11" of MG (Scott *et al.*, 1994).

Detecting Immune Response to Vaccination

Although, there is a lack of correlation between levels of circulating antibodies and protection against MG, serology is still commonly used to determine whether a vaccine has elicited immune response or not. More importantly, serology is also used to establish that a flock has not been infected prior to vaccination. The procedures used are the same as those used in diagnosis including RSPAT, HI and ELISA (Kleven, 1994).

However, one of the disadvantages of currently available vaccines is that there is no convenient serological technique to distinguish precisely between vaccinated or naturally infected flocks (Whithear and Browning, 1994).

Vaccination

The objectives of vaccination are:

1. To prevent the egg production losses and respiratory disease associated with MG infections in layer and breeder flocks.
2. To prevent the egg production losses, respiratory disease and leg problems associated with MS infections in layer and breeder flocks.
3. To reduce medication costs involved in the treatment of mycoplasma infected flocks.
4. To reduce the vertical transmission rates of MG from breeders to progeny chicks.
5. To reduce the build up of infectious reservoirs of mycoplasma.

Layer Mycoplasma Inactivated Vaccination

Programme – 1

This is inactivated bacterin vaccination programme designed for use in commercial layer flocks which are raised as mycoplasma free pullets and expected to become infected duing the lay.

One inactivated bacterin vaccination prior to the onset of egg production (or prior to mooving to the layer facility) reduces loss in egg production due to mycoplasma infection. This should be done 4 weeks before the onset of egg production *i.e.* at 12th week of age of birds. The inactivated bacterin vaccination should be given intramuscularly or subcutaneously.

Precautions

Mycoplasma bacterins can cause post vaccinal swellings at the site of inoculation, stiff neck and swollen heads. These reactions are more likely to occur when the vaccine is placed too close to the head or into the muscles of neck. The residue and tissue reaction from vaccination with oil-adjuvented inactivated bacterins may be detectable for many weeks post-vaccination.

Bacterins don't prevent tracheal infections, but minimize tracheal lesions, prevent the spread of the infection to the lower respiratory tract and oviduct reducing the vertical transmission of MG significantly. Mycoplasma free flock should not be vaccinated. Immunity against MG doesn't protect against MS or vice-versa. Vaccinated flocks may test positive on the RSPAT, HI and ELISA tests.

Breeder Mycoplasma Inactivated Vaccination

Programme – 1

This inactivated bacterin vaccination programme is for use in breeder flocks which are raised as mycoplasma free pullets and expected to become infected during the lay.

In breeders, two doses of inactivated bacterin vaccination are required to decrease the vertical transmssion of MG organisms to the progeny chicks. The bacterin vaccinations should be at least 4 weeks apart to allow the optimum immune response to occur. The second bacterin dose should be given at least 4 weeks prior to the onset of egg production (*i.e.* at 12th week of age) or exposure to field challenge. It is already clear that the first bacterin dose should be administered at the age of 8th week and second bacterin dose should be administered at the age of 12th week of breeders, as birds generally start egg laying by 16th week of age in well managed flocks.

The precautions which should be concidered are as per the description in layer mycoplasma vaccination. In breeders, one bacterin vaccination does not effectively reduce the rate of vertical transmission, so two doses are required.

Layer/Breeder Mycoplasma Inactivated Vaccination

Programme – 1

This inactivated bacterin vaccination programme is for use in layer and breeder flocks which become infected with mycoplasma as pullets.

Two inactivated bacterin vaccinations should be given to stimulate adequate immune protection. The first vaccination is given 3-4 weeks prior to the expected field infection. The age of field challenge is estimated by serology or isolation of the mycoplasmafrom pullet flocks. Because actual time of infection is difficult to determine, the first vaccination is often given at 7 -14 days of age. The second bacterin vaccination should be given at least 4 weeks prior to the onset of egg production *i.e.* at 12th week of age of birds in well managed flocks.

Precautions are same as in layer mycoplasma vaccination.

Programme – 2

This inactivated bacterin vaccination programme is for use in layer and breeder flocks which have come from parent stock known to be infected with mycoplasma.

Birds should be treated with a suitable antimicrobial agent for the first 10 days of age. The purpose of medication is to limit the horizontal spread of the mycoplasma organisms from the vertically infected chicks to uninfected susceptible chicks. The first inactivated bacterin vaccination should be given at 7–10 days of age to establish an early active immune response. The second inactivated bacterin vaccination is given 4 weeks prior to the onset of egg production *i.e.* at 12th week of age of birds in well managed flocks.

In breeders, one bacterin vaccination doesn't effectively reduce the rate of vertical transmission of mycoplasma, hence two bacterin vaccination doses are recommended.

Precautions are same as in layer mycoplasma vaccination.

Layer/Breeder Mycoplasma gallisepticum Live Vaccination

This live MG vaccination programme is for commercial layer and breeder flocks which become infected as pullets or during lay.

The long term use of some live MG vaccines (F strain, 6/85 strain, ts-11 strain) may displace the field strain in the house environment. Vaccination of birds with live vaccines may not prevent infection by the field strain. Use of live vaccine in breeder flocks should be done with extreme caution. Vaccinated flocks will test positive on the RSPAT, HI and ELISA tests. Milder live strain containing live MG vaccines don't elicite strong serologic response. Flock vaccinated with live vaccines may be positive for MG when cultured. Vaccination of flocks in lay may be associated with decrease in egg production. It is recommended that no antibiotic medication should be administered with in 7 days of live MG vaccines application to flocks. Turkeys or broilers should not be exposed to "F strain" or to chickens vaccinated with "F strain" of MG as live vaccine. Milder strains (6/85 or ts-11) of live vaccines are reported to be safe for turkeys. However, mild live MG vaccines don't spread well among birds. No other live vaccine should be administered with in 7 days of vaccination with live MG vaccines. The live vaccines may be given as intra-ocular or in drinking water or as fine spray as per the strain involved in the vaccine being applied.

Precaution

Live ND (New castle disease), IB (infectious bronchitis) and ILT (infectious laryngotracheitis) vaccines given to laying birds may result in more severe reactions after live mycoplasma vaccination.

Layer/Breeder Mycoplasma Synoviae Live Vaccination

The live MS vaccination is for use in layers and breeders to prevent synovitis, respiratory signs and associated lameness of birds.

The strain (MS – H) of M. synoviae is an attenuated strain, which is currently undergoing evaluation. This MS – H strain live vaccine is used only in the flocks or areas where this disease is diagnosed properly. This live vaccine is administered and distributed in the same way as ts-11 strain vaccine of MG. MS–H vaccine is commonly used in to the alternate eyes at the same time as vaccinated with to 11 strain vaccine of MG.

Broiler MG/MS Vaccination

Vaccination of broilers for MG or MS is not practiced. Eradication of MG/MS infections in broiler breeders is the best method of control of mycoplasma in broilers.

REFERENCES

Adler, H. E. (1976). Immunological response to *Mycoplasma gallisepticum*. *Theriogenology*, **6**, 87-91.

Ando, K., Matsui, K., Sato, S., Yoshida, I., Kato, K. and Kuniyasu, C. (1965). Evaluation of antigenicity of the agglutination antigen for avian respiratory mycoplasmosis and availability of the antigen for the field test. *Natl. Inst. Anim. Hlth Q.*, **5**, 1 3 - 1 9.

Animal and Plant Health Inspection Service (APHIS) (1997). National Poultry Improvement Plan guidelines and regulations. Vol. APHIS-91-55-038. United States Department of Agriculture, Beltsville, Maryland, 105 pp.

Avakian, A. P. and Kleven, S. H. (1990). The humoral immune response of chickens to *Mycoplasma gallisepticum* and Mycoplasma synoviae studied by immunoblotting. *Vet. Microbiol.*, **24 (2)**, 155-169.

Avakian, A. P. and Ley, D. H. (1993). Protective immune response to *Mycoplasma gallisepticum* demonstrated in respiratory-tract washings from *M. gallisepticum* infected chickens. *Avian Dis.*, **37 (3)**, 697-705.

Bradbury, J. M., Yavari, C. A. and Giles, C. J. (1994). *In vitro* evaluation of various antimicrobials against Mycoplasmagallisepticum and Mycoplasma synoviae by the micro-broth method, and comparison with a commercially-prepared test system. *Avian Pathol.*, **23**, 105-115.

Bencina, D., Mrzel, L., Svetlin, A., Dorrer, D. and Tadina-Jaksic T. (1991). Reactions of chicken biliary immunoglobulin A with avian mycoplasmas. *Avian Pathol.*, **20**, 303-313.

Bencina, D., Tadina, T. and Dorrer, D. (1988). Natural infection of geese with *Mycoplasma gallisepticum* and Mycoplasma synoviae and egg transmission of the mycoplasmas. *Avian Pathol.*, **17**, 925-928.

Brown, M. B., Stoll, M. L., Scasserra, A. E. and Butcher, G. D. (1991). Detection of antibodies to *Mycoplasma gallisepticum* in egg yolk versus serum samples. *J. Clin. Microbiol.*, **29 (12)**, 2901-2903.

Charlton, B. R., Bickford, A. A., Walker, R. L. and Yamamoto, R. (1999). Complementary randomly amplified polymorphic DNA (RAPD) analysis patterns and primer sets to differentiate *Mycoplasma gallisepticum* strains. J. vet. diagn. Invest, **11 (2)**, 158-161.

Christensen, N. H., Yavari, C. A., McBain, A. J. and Bradbury, J. M. (1994). Investigations into the survival of *Mycoplasma gallisepticum*, Mycoplasma synoviae, and Mycoplasma iowae on materials found in the poultry house environment. *Avian Pathol.*, **23**, 127-143.

Droual, R., Bickford, A. A. and Cutler, G. J. (1993). Local reaction and serological response in commercial layer chickens injected intramuscularly in the leg with oiladjuvanted *Mycoplasma gallisepticum* bacterin. *Avian Dis.*, **37**, 1001-1008.

Evans, R. D. and Hafez, Y. S. (1992). Evaluation of a *Mycoplasma gallisepticum* strain exhibiting reduced virulence for prevention and control of poultry mycoplasmosis. *Avian Dis.,* **36,** 197-201.

Fabricant, J. (1975). - Immunization of chickens against *Mycoplasma gallisepticum* infection. *Am. J. Vet. Res.,* **36 (4 Pt 2),** 566-567.

Fan, H. H., Kleven, S. H. and Jackwood, M. W. (1995). Application of polymerase chain reaction with arbitrary primers to strain identification of *Mycoplasma gallisepticum. Avian Dis.,* **39 (4),** 729-735.

Gallazzi, D., Enice, F., Fabris, G. and Mandelli, G. (1985). Prove di vaccinazione in campo contro le infezioni aviarie da *Mycoplasma gallisepticum. Clinica Vet., Milano,* **108** (2), 115-121.

Glisson, J. R., Dawe, J. F. and Kleven, S. H. (1984). The effect of oil-emulsion vaccines on the occurrence on nonspecific plate agglutination reactions for *Mycoplasma gallisepticum* and M. synoviae. *Avian Dis.,* **28 (2),** 397-405.

Glisson, J. R. and Kleven, S. H. (1984). – *Mycoplasma gallisepticum* vaccination: effects on egg transmission and egg production. *Avian Dis.,* **28 (2),** 406-415.

Hidaka, S., Terazawa, G., Nakaim, M., Hotta, H., Seko, M., Yamamoto, M., Tanaka, R. And Konishi, T. (1987). Removal of nonspecific plate agglutination reaction for *Mycoplasma gallisepticum* and Mycoplasma synoviae in chicken serum vaccinated with IBD vaccines [in Japanese]. *J. Jpn. Soc. Poult. Dis.,* **23,** 3 1 - 3 6.

Hildebrand, D. G., Page, D. E. and Berg, J. R. (1983). *Mycoplasma gallisepticum* (MG) - laboratory and field studies evaluating the safety and efficacy of an inactivated MG bacterin. *Avian Dis.,* **27,** 792-802.

Kempf, I. (1998). DNA amplification methods for diagnosis and epidemiological investigations of avian mycoplasmosis. *Avian Pathol.,* **27,** 7-14.

Kempf, I., Gesbert, F., Guittet, M. and Bennejean, G. (1994). *Mycoplasma gallisepticum* infection in drug-treated chickens: comparison of diagnostic methods including polymerase chain reaction. *J. Vet. Med.,* B, **41,** 597-602.

Khan, M. I., Kirkpatrick, B. C. and Yamamoto, R. (1989). *Mycoplasma gallisepticum* species and strain specific recombinant DNA probes. *Avian Pathol.,* **18,** 135-146.

Kleven, S. H. (1975). Antibody response to avian mycoplasmas. *Am. J. Vet. Res.,* **36,** 563-565.

Kleven, S. H. (1985). Tracheal populations of *Mycoplasma gallisepticum* after challenge of bacterin-vaccinated chickens. *Avian Dis.,* **29** (4), 1012-1017.

Kleven, S. H. (1994). Avian mycoplasmas. In: *Mycoplasmosis in Animals: Laboratory Diagnosis* (H.W. Whitford, R.F. Rosenbusch and L.H. Lauerman, eds). Iowa State University Press, Ames, Iowa, 31-38.

Kleven, S.H. (1998). - Mycoplasma in the etiology of multifactorial respiratory diseases. *Poult Sci.,* **77,** 1146-1149.

Kleven, S. H., Jordan, F. T. W. and Bradbury, J. M. (1996). Avian mycoplasmosis (*Mycoplasma gallisepticum*). In: *Manual of Standards for Diagnostic Tests and Vaccines*, 3rd Ed. Office International des Epizooties, Paris, 512-521.

Kleven, S. H., Morrow, C. J. and Whithear, K. G. (1988). Comparison of *Mycoplasma gallisepticum* strains by hemagglutination-inhibition and restriction endonuclease analysis. *Avian Dis.*, **32**, 731-741.

Kleven, S. H., Rowland, G. N. and Olson, N. O. (1991). Mycoplasma synoviae infection. In: *Diseases of Poultry* (B.W. Calnek, H.J. Barnes, C.W. Beard, W.M. Reid and H.W. Yoder Jr, eds). Iowa State University Press, Ames, Iowa, 223-231.

Krause, D. C. (1996). *Mycoplasma pneumoniae* cytadherence: unravelling the tie that binds. Molec. *Microbiol.*, **20 (2)**, 247-253.

Kume, K., Kawakubo, Y., Morita, C., Hayatsu, E. and Yoshioka, M. (1977). Experimentally induced synovitis of chickens with Mycoplasma synoviae: effects of bursectomy and thymectomy on course of the infection for the first four weeks. *Am. J. Vet. Res.*, **38**, 1595-1600.

Kuniyasu, C. and Ando, K. (1966). Studies on the hemagglutination-inhibition test for *Mycoplasma gallisepticum* infection of chickens. *Natl Inst. Anim. Hlth Q.*, **6**, 136-143.

Lam, K. M. and Lin, W. (1984). Resistance of chickens immunized against *Mycoplasma gallisepticum* is mediated by bursal dependent lymphoid cells. *Vet. Microbiol.*, **9(5)**, 509-514.

Lauerman, L. H. (1998). Manual on: nucleic acid amplification assays for diagnosis of animal diseases. American Association of Veterinary Laboratory Diagnosticians, Turlock, California, 166 pp.

Levisohn, S., Glisson, J. R. and Kleven, S. H. (1985). in ovo pathogenicity of *Mycoplasma gallisepticum* strains in the presence and absence of maternal antibody. *Avian Dis.*, **29**, 188-197.

Levisohn, S. and Kleven, S. H. (1981). Vaccination of chickens with nonpathogenic *Mycoplasma gallisepticum* as a means for displacement of pathogenic strains. *Isr. J. Med. Sci.*, **17**, 669-673.

Ley, D. H., Mclaren, J. M., Miles, A. M., Barnes, H. J., Heinsmiller, S. and Franz, G. (1997). Transmissibility of live *Mycoplasma gallisepticum* vaccine strains ts-11 and 6/85 from vaccinated layer pullets to sentinel poultry. *Avian Dis.*, **41**, 186-193.

Ley, D. H. and Yoder, H. W., Jr (1997). *Mycoplasma gallisepticum* infection. In *Diseases of Poultry*, 10th Ed. (B. W. Calnek with H. J. Barnes, C. W. Beard, L. R. McDougald and Y. M. Saif, eds). Mosby-Wolfe, London, 194-207.

Lin, M. Y. and Kleven, S. H. (1982). Pathogenicity of two strains of *Mycoplasma gallisepticum* in turkeys. *Avian Dis.*, **26**, 360-364.

Minamimoto, S., Suzuki, K., Tanaka, K., Otaki, K. and Murayama, J. (1968). Eradication of *Mycoplasma gallisepticum* infection in a chicken flock on a breeding farm. *Natl Inst. Anim. Hlth Q.*, **8**, 164-168.

Mohammed, H. O., Carpenter, T. E. and Yamamoto, R. (1987). Economic impact of *Mycoplasma gallisepticum* and M. synoviae in commercial layer flocks. *Avian Dis.*, **31 (3)**, 477-482.

Murayama, J. (1984). Development of thermal regulation systems for preincubation heat treatment of hatching eggs to eliminate Mycoplasma infections in breeding chicken flocks [in Japanese]. *Anim. Hus.*, **38**, 527-530.

Murayama, J., Fueki, M., Onojima, M., Iwasa, N. and Oda, S. (1977). Eradication of Mycoplasma synoviae from breeding chicken flocks by means of preincubation heat treatment of hatching eggs [in Japanese]. *Anim. Hus.*, **31**, 1123-1125.

Nonomura, I., Sato, S., Syoya, S., Shimizu, F. and Horiuchi, T. (1971). Multiplication of *Mycoplasma gallisepticum* and Newcastle disease virus B1 strain in the respiratory tract of chickens. *Natl. Inst. Anim. Hlth Q.*, **11**, 1-10.

Nunoya, T., Kanai, K., Yagihashi, T., Hoshi, S., Shibuya, K. and Tajima, M. (1997). Natural case of salpingitis apparently caused by *Mycoplasma gallisepticum* in chickens. *Avian Pathol*, 26 (2), 391-398.

Nunoya, T., Yagihashi, T., Tajima, M. and Nagasawa, Y. (1995). Occurence of keraconjunctivitis apparently caused by *Mycoplasma gallisepticum* in layer chickens. *Vet. Pathol.*, **32**, 11-18.

Razin, S. (1992). Peculiar properties of mycoplasmas: the smallest self-replicating prokaryotes. *FEMS Microbiol Lett.*, **100**, 423-432.

Razin, S., Yogev, D. and Naot, Y. (1998). Molecular biology and pathogenesis of mycoplasmas. *Microbiol. Mol. Biol. Rev.*, 62(4), 1094-1156.

Roberts, D. H. and McDaniel, J. W. (1967). Mechanism of egg transmission of *Mycoplasma gallisepticum*. *J. Comp. Pathol.*, 77, 439-442.

Rose, M. E. and Orlans, E. (1981). Immunoglobulins in the egg, embryo and young chick. *Dev. Comp. Immunol.*, **5**, 15-20.

Sato, S. (1976). Mycoplasma synoviae infection in chickens. *Jpn Agric. Res. Q.*, **10**, 94-100.

Scott, P. C., Jones, J., Morrow, C. J., Ley, D. H. and Whithear, K. G. (1994). Experiences with a live attenuated Mycoplasma synoviae vaccine. In: *Proc. 43rd Western Poultry Disease Conference* (M.M. Jensen, ed.). Sacramento, 27 February-1 March, 97-98.

Talkington, F. D. and Kleven, S. H. (1985). Evaluation of protection against colonization of the chicken trachea following administration of *Mycoplasma gallisepticum* bacterin. *Avian Dis.*, **29**, 998-1003.

Vardaman, T. H., Landreth, K., Whatley, S., Dreesen, L. J. and Glick, B. (1973). Resistance to Mycoplasma synoviae is bursal dependent. *Infect. and Immunity*, **8**, 674-676.

Whithear, K. G. (1996). Control of avian mycoplasmoses by vaccination. In Animal mycoplasmoses and control. Nicolet, ed.). *Rev. Sci. Tech. Off Int. Epiz.*, **15 (4)**, 1527-1553.

Whithear, K. G. and Browning, G. F. (1994). Mycoplasmas and the management of broiler breeders. In: *Proc. 43rd Western Poultry Disease Conference* (M.M. Jensen, ed.). Sacramento, 27 February-1 March, 59-60.

Whithear, K. G., Ghiocase, E., Markhaam, P. F. and Marks, D. (1990). Examination of *Mycoplasma gallisepticum* isolates from chickens with respiratory disease in a commercial flock vaccinated with a living *M. gallisepticum* vaccine. *Aust. Vet. J.*, **67 (12),** 459-460.

Whithear, K. G., Soeripto, Harrigan, K. E. and Ghiocas, E. (1990). Immunogenicity of a temperature sensitive mutant *Mycoplasma gallisepticum* vaccine. *Aust. Vet. J.*, **67** (5), 168-174.

Yagihashi, T. and Tajima, M. (1986). Antibody responses in sera and respiratory secretions from chickens infected with *Mycoplasma gallisepticum*. *Avian Dis.*, **30,** 543-550.

Yamamoto, H., Watanabe, H., Sato, G. and Mikami, T. (1975). Identification of immunoglobulins in chicken eggs and their antibody activity. *Jpn. J. Vet. Res.*, **23,** 131-140.

Yoder, H. W. JR (1970). Preincubation heat treatment of chicken hatching eggs to inactivate Mycoplasma. *Avian Dis.*, **14,** 75-86.

Yoder, H. W. JR (1991). *Mycoplasma gallisepticum* infection. In Diseases of poultry (B.W. Calnek, C.W. Beard, H.J. Barnes, W.M. Reid and H.W. Yoder Jr, eds). Iowa State University Press, Ames, Iowa, 198-212.

Yoder, H. W. JR, Hopkins, S. R. and Mitchell, B. W. (1984). Evaluation of inactivated *Mycoplasma gallisepticum* oil emulsion bacterins for protection against air sacculitis in broilers. *Avian Dis.*, **28,** 224-234.

Zanella, A., Martino, P. A., Pratelli, A. and Stonfer, M. (1998). Development of antibiotic resistance in *Mycoplasma gallisepticum in vitro*. *Avian Pathol*, **27,** 591-596.

Index